Builders' Costs for 100 Best-Selling Home Plans

The First and Only Resource that Combines Summary Cost Estimates with Designs

Produced by the leading providers of building cost data and home plans, R.S. Means Co., Inc. and Home Planners, Inc.

RSMeans.

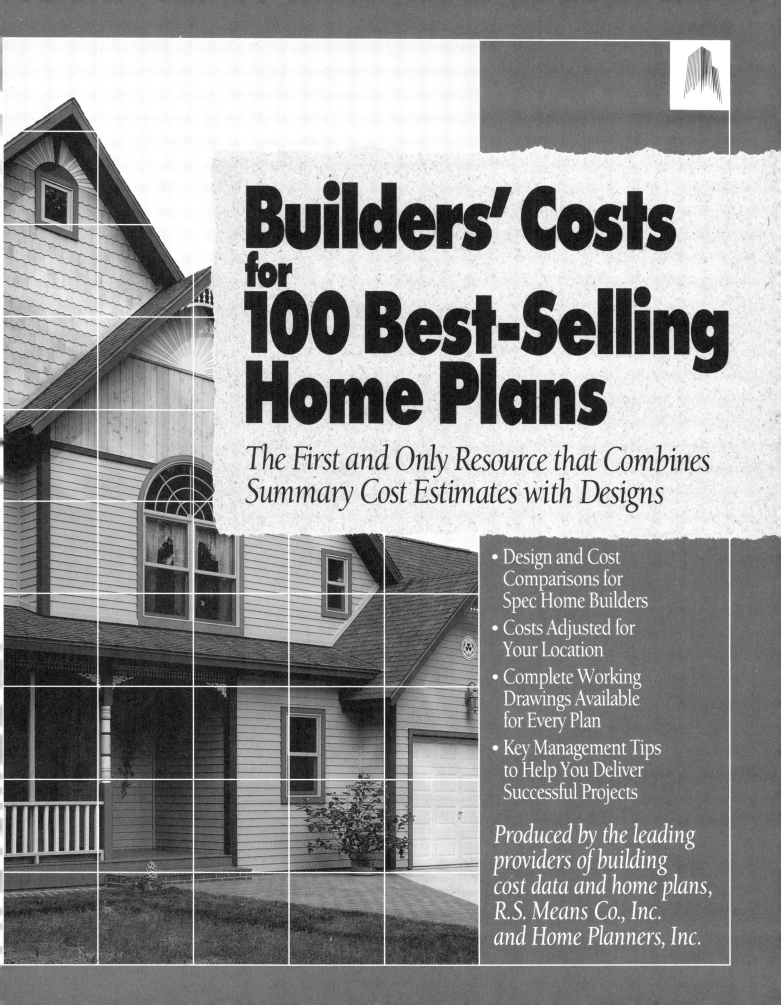

Builders' Costs
for
100 Best-Selling Home Plans

*The First and Only Resource that Combines
Summary Cost Estimates with Designs*

- Design and Cost Comparisons for Spec Home Builders
- Costs Adjusted for Your Location
- Complete Working Drawings Available for Every Plan
- Key Management Tips to Help You Deliver Successful Projects

Produced by the leading providers of building cost data and home plans, R.S. Means Co., Inc. and Home Planners, Inc.

RS Means®

Copyright 1995

R.S. MEANS COMPANY, INC.
CONSTRUCTION PUBLISHERS & CONSULTANTS

100 Construction Plaza
P.O. Box 800
Kingston, MA 02364-0800
(617) 585-7880

The editors for this book were Mary Greene, Patricia L. Jackson, PE, Mark Kaplan, Jr. and Robert Mewis. The production manager was Helen Marcella; the production coordinator was Marion Schofield. Composition was supervised by Karen O'Brien. Technical assistance provided by Gary Hoitt, Scott Nichols and Kathryn Rodriguez. The book and cover were designed by Norman R. Forgit.

10   9   8   7   6   5   4   3

Library of Congress Cataloging in Publication Data

ISBN 0-87629-356-9

Table of Contents

Foreword

Residential contractors, whether building speculation or custom homes, have an ongoing need for design ideas and solutions. They also need to be able to compare the costs of various designs and design features in an easy and efficient way. This new publication from Home Planners, Inc. and R.S. Means Company, Inc. was created to meet those needs. It also serves as a useful tool for qualifying clients' budgets and expectations, thereby saving everyone time and money.

The book contains 100 up-to-date plans created by the Home Planners design team, with summary estimates for each plan prepared by R.S. Means Co. Various styles are represented, including Traditional, Cape Cod, Colonial, Southwestern, Contemporary, and Ranch, ranging in size from 1300 to 4200 square feet. Of the thousands of plans available from Home Planners, these were selected from best-sellers that offered the following features: an eye-pleasing front elevation, and a relatively simple and efficient shape and floor plan layout. These designs are particularly applicable to speculation building due to their winning combination of economy and market appeal. Offered are styles for all regions of the U.S. and Canada. Location factors allow you to easily adjust cost estimates to your specific location for 927 cities and towns.

While these plans all adhere to accepted building standards, they may have to be modified to meet local codes and site conditions, and such modifications will affect the total cost. Materials priced in these estimates are of a moderate quality level; luxury-grade materials would, of course, increase the overall building cost. Basic excavation requirements are covered, while site work and landscaping are not, since these items are site- and climate-specific. A complete set of plans is available for each of the houses illustrated and estimated in this book. (See order information at the back of the book.)

Also included in the book are "Builder's Management Tools," a review of information on procuring building permits, using forms to manage a project, basic guidelines for selecting building sites, and more.

For 52 years, R.S. Means has been the leading provider of cost data to professionals in every facet of the construction industry. Means currently publishes 24 annual cost data books covering a wide range of construction types and specialties. In addition, the company offers over 60 reference books which provide business, financial and project management guidance. Means also offers seminars, electronic data and consulting services.

Home Planners, Inc., founded in 1946, is the preeminent provider of home plan books and blueprints. The company has developed a portfolio of working drawings for over 3,000 different homes. Home Planners also publishes 23 books of home designs, distributed in bookstores, libraries, home improvement centers and other stores, as well as its own magazine, *Home Planner,* published eight times a year. Home Planners' plans are shown in national magazines as well, such as *House Beautiful's* "Houses and Plans" specials and *Home Building* and *Colonial Homes* as well as in various publications of *Better Homes and Gardens.*

Section 1
Management Tools for Builders

The following pages include project and business management information, together with a few forms. Some of these sections cover key topics for novice builders, such as interpreting plans and working with the building department. However, even experienced builders will find it worthwhile to review the topics that apply to the parts of their work that they feel they may be able to do better. Whatever your level of experience, Section I is intended to clarify some basic issues, and to stimulate the pursuit of the details you need on the subjects most important to you. A list of recommended references is included at the back of this book.

Section I includes:
- Qualities of the Successful Home Builder
- Plan Reading Review
- Site Selection Overview
- Selecting a Spec Building Plan
- Procuring the Building Permit
- Forms for Home Builders
- Safety Tips

Qualities of the Successful Home Builder

Sound construction knowledge is clearly a basic qualification of the successful home builder, whether the focus is on spec or custom homes. Taken alone, however, it is not enough. An experienced tradesman may possess the skill to actually construct a project, yet still fall short in other important areas. Motivation, leadership, and organizational and decision-making ability are other key qualifications, along with a talent for public relations and marketing.

Public relations talent is valuable in daily negotiations and communications with building departments, potential buyers, realtors, subcontractors, suppliers, and others. Motivation provides the spirit to accept a challenge, and leadership lies in the natural and comfortable acceptance of the managing role. The many diverse contributors to a project look to its manager for information, dates, directions, and approvals. Decision-making routinely includes budgeting for, selecting, and ordering materials; identifying, qualifying, and hiring subcontractors; planning work activities; and choosing the most efficient construction methods and equipment. Added to these purely construction-related functions are a number of major financing and marketing tasks and responsibilities.

The contractor's experience, ability, and good judgment are the foundation of leadership and success in this industry.

The ability to plan and organize is essential for obvious reasons. Most of the planning must be accomplished by the contractor personally. It is a key element in these tasks:

- Researching and selecting the most suitable locations for new homes.

- Negotiating and purchasing the land on which the home is to be built.

- Selecting, studying and where necessary, revising the drawings and specifications (and having those revisions approved), reviewing the subcontractors' bid proposals in detail, and being thoroughly acquainted with every project, while staying current with the best sources of materials and labor.

- Arranging financing.

- Drawing up a construction progress schedule; then using this schedule as a tool in starting subcontractors and suppliers as early as possible and directing each of them to the shortest possible completion times. With advance warning and cooperation in scheduling, subcontractors can assist each other to mutual advantage.

- Estimating construction project costs, from an organized and thorough quantity takeoff to pricing materials, as well as seeking out and learning how to work with the best suppliers and subcontractors—responsible craftsmen and bidders who can give you a reasonable price.

- Continuously contacting subcontractors and suppliers to maintain schedule requirements, making adjustments to accommodate unexpected developments such as weather-related or material delivery delays, and whatever else it takes to keep the project on target and on budget.

- Marketing and selling the home, including advertising, listing with a realtor, and attending the closing meeting.

- Arranging the final building inspection, obtaining a certificate of occupancy, performing a walk-through with the owners, and correcting any punch list items.

Despite the opportunities in the home building industry, many contractors go out of business before long because while they are meticulous craftsmen, they lack the necessary business skills. Unfortunately, many do not realize that the knowledge and experience they need are within reach—through the benefits offered by membership in national and local industry associations, seminars and training, and reference books written by successful professionals. All of these information sources are tools, like a hammer and saw, that when used effectively, will increase productivity and get the job done.

Plan Reading Review

Site Plan

Due to the site-specific requirements of each home and location, stock plan companies do not offer site plans as part of the plan packages they sell. Site plans are typically drawn up by an architect or civil engineer. The main purpose of the basic site plan is to locate the structure within the confines of the building lot. Even the most basic of site plans clearly establishes the building's dimensions, usually by the foundation's size and the distances to the respective property lines. The latter, called the *setback dimensions,* are shown in feet and hundredths of a foot, versus feet and inches on the architectural drawings. For example, the architectural dimension of 22'-6" would be 22.50' on a site plan.

To obtain the most information about the site in preparation for the site design, a *site survey* is generally performed by a registered land surveyor. The land surveyor records special conditions present on the building lot. This includes locating existing natural features such as trees or water, as well as man-made improvements such as walks, paving, fences, or other structures. The new site plan shows how the existing features are kept, modified, or removed to accommodate the new design.

Another chief purpose of the site plan is to show the unique surface conditions, or *topography,* of the lot.

The topography of a particular lot may be shown right on the site plan. For projects in which the topography must be shown separately for clarity, a grading plan is used. The topographical information includes changes in the elevation of the lot such as slopes, hills, valleys, and other variations in the surface. These changes in the surface conditions are shown on a site plan by means of a *contour,* which is a line connecting points of equal elevation. An elevation is a distance above or below a known point of reference, called a *datum.* This datum could be sea level, or an arbitrary plane of reference established for the particular building.

The existing contour is typically shown as a dashed line, with the new or proposed contour shown as a solid line. Both may be labeled with the elevation of the contour in the form of a whole number. The spacing between the contour lines is at a constant vertical increment, or interval. The typical interval is five feet, but intervals of one foot are not uncommon for site plans requiring greater detail (or where the change in elevation is more dramatic).

Two important characteristics of the contour need to be observed when reading a site plan.

- Contours are continuous, and frequently enclose large areas in comparison to the size of the building lot. For this reason, contours are often drawn from one edge to the other edge of the site plan.

- Contours do not intersect or merge together. The only exception to this rule is in the case of a vertical wall or plane. For example, a retaining wall shown in plan view would show two contours touching, and a cliff that overhangs would be the intersection of two contours.

A known elevation on the site for use as a reference point during construction is called a *benchmark.* The benchmark is established in reference to the datum and is commonly noted on the site print with a physical description and its elevation relative to the datum. For example: "Northeast corner of catch basin rim – Elev. 102.34'" might be a typical benchmark found on a site plan. When individual elevations, or *grades,* are required for other site features, they are noted with a " + " and the grade. Grades vary from contours in that a grade has accuracy to two decimal places, whereas a contour is expressed as a whole number.

Some site plans include a small map, called a *locus,* showing the general location of the property with respect to local highways, routes, and roads. Sophisticated site plans showing utilities and drainage services often require a legend. The legend is similar to that on the architectural drawings, listing the different symbols and abbreviations found in the particular group of site plans.

The "north arrow" clearly shows the direction of magnetic north as a reference for naming particular sides or areas of the project. In addition, the surveyor labels the property lines in accordance with the directions normally found on a compass. This reference, in the form of an angle and its corresponding distance, is called the *bearing* of a line. The bearings of the encompassing property lines are often the legal description of the building lot.

Architectural Drawings

The core drawings include the *architectural drawings*. Architectural drawings are typically numbered sequentially and may include basement or ground-floor plans, upper-level floor plans, exterior elevations, sections, interior elevations, details, and window, door, and room finish schedules.

Plan View   One of the architect's tools is the *plan view*. The most common type of plan view is the *floor plan*. The function of the architectural floor plan is to show the use of space. Floor plans identify the locations of rooms, stairs, means of egress, and where traffic will flow from room to room. Floor plans show the locations of major features such as windows, partitions, interior doorways, and built-ins such as cabinetry and bookcases.

Architectural plans should be dimensioned to show actual length and width, thereby allowing the reader to calculate areas. Dimensions should be accurate, clear, and complete, showing both exterior and interior measurements of the space.

Architectural plans include notes that further define and provide information about a particular segment of work, or reference another drawing. In residential and light commercial construction, a separate set of specifications is not always issued, depending on the size of the projects and financial constraints. In many cases, it is assumed that the notes will suffice.

The plan view approach is not limited to floors, but can be used in other drawings such as roof or framing plans to provide the same perspective. *Partial floor plans* show an enlarged view of smaller areas such as bathrooms, bedrooms, kitchens, and stairs.

Perspective View or Rendering

Many potential customers have problems visualizing what a home will look like on a lot. Mechanically drawn floor plans and elevations are sometimes too antiseptic for this purpose. A perspective or rendering is drawn to show the home as it will actually be seen—all vertical and horizontal lines extended to a vanishing point. Many times a perspective, or rendering, is better than a photograph because it allows the draftsman to include attractive features, such as lawn, trees, shrubs and driveways to be added later.

Exterior Elevations   An important part of the architectural drawings are the *exterior elevations*. They provide a pictorial view of the exterior of the structure. Unlike a floor plan, an elevation is a representation of the exterior wall viewed perpendicular to both the vertical and horizontal planes. Elevations are not in perspective view, which makes it difficult to determine depths or changes in direction without looking at all the elevations.

Exterior elevations may be titled based on their location with respect to the headings of a compass (North Elevation, South Elevation, East Elevation, and West Elevation), or may be titled Front Elevation, Rear Elevation, Right and Left Side Elevations. The scale of the elevation is noted either in the title block or under the title of the elevation.

The function of the elevation is to provide a clear depiction of exterior doors, windows, and the overall aesthetic appearance and style of the building, often using numbers or letters in circles to identify the doors and windows as listed in the door and window schedule. In addition, elevations show the surface materials of the exterior wall, and any changes in the surface materials within the plane of the elevation.

While the floor plan shows measurements in a horizontal plane, elevations provide measurements in a vertical plane, with respect to a horizontal plane. These dimensions provide a vertical location of floor-to-floor heights, window sill or head heights, floor-to-plate heights, roof heights, or a variety of dimensions from a fixed horizontal surface. By using these measurements, the reader can calculate quantities of materials needed. Occasionally the elevation dimensions are given as decimals (10.5′ as opposed to 10′-6″). Along with the dimensions on the elevations, notes are often included for clarification.

Building Section The building section, commonly referred to as the *section,* is a "vertical slice" or cut through a particular part of the building. It offers a view through a part of the structure. Several different sections may be incorporated into the drawings. Sections taken from a plan view are called *cross-sections;* those taken from an elevation are referred to as *longitudinal sections,* or simply *wall sections.* Wall sections offer the reader an exposed view of the components and the arrangement within the wall itself. Using sections in conjunction with floor plans and elevations, one begins to understand how the building goes together.

The scale of the section may be different than previously encountered in the plan and elevations. A general rule is that the scale increases as the section in question becomes smaller. For example, typical architectural drawings use 1/4″ = 1′ for plan views, but increase the scale to 3/8″ = 1′ or 1/2″ = 1′ for building sections. A typical wall section should provide dimensions vertically, and any relative dimensions horizontally, such as wall thickness, setbacks, overhangs, or similar changes in the vertical plane.

Details For greater clarification and understanding, often certain areas of a floor plan, elevation, or a particular part of a drawing may need enhancement. In this case, these sections are drawn to a larger scale and are referred to as *details.* Details can be found either on the sheet where they are first referenced, or grouped together on a separate sheet included in the set of drawings. Details are one of the most important sources of information available to the contractor and estimator. Often the detail is shown in larger scale to provide additional space for recording dimensions and notes. Details are not limited to architectural drawings but can be used in structural and site plans and, to a lesser degree, in mechanical or electrical plans.

Schedules In an effort to keep the drawings from becoming cluttered with too much printed information or details, architects have devised a system to incorporate information pertaining to a similar group of items. For example, listings of doors, windows, the finishes of the rooms, columns, trusses, and lighting or plumbing fixtures can be set in an easy-to-read table. These tables are called *schedules.*

Each door is listed on the schedule by number and provides information such as size and type, thickness, frame material, and hardware. In addition, the door schedule often states any specific instructions for an individual door, such as fire ratings,

undercutting, weather-stripping, or vision panels. The "Remarks" portion of the schedule offers the architect the opportunity to list any nonstandard requirements for the door.

It is easy to see how the architectural drawings have become the focal point of a set of working drawings. But the architectural drawings alone do not provide sufficient information to construct a building. There are many other systems that are part of the completed structure that the architectural drawings do not address. These include the structural components of the building, the plumbing, heating, and electrical systems, and the site plan for the structure.

Mechanical Drawings These sheets in an architectural drawing typically show how the heating, air-conditioning and plumbing systems are laid out and constructed for the house. Because these systems and the appliances and materials they use are highly dependent upon local conditions and regulations, a local mechanical engineer or HVAC or plumbing contractor often supplies the drawings and calculations for these sheets. If drawings are purchased from a stock plan company selling nationally, these pages are not usually included in the package. However, in the case of Home Planners, Inc., generic drawings can be purchased which show typical plumbing and HVAC components and layouts.

Generally, the mechanical drawings follow the structural drawings and comprise the different mechanical systems of the building. Specifically, they include the plumbing drawings, the heating, ventilating, and air-conditioning drawings, and the fire protection drawings. Most of the work shown on these three types of drawings is in plan view form. Because of the diagrammatic nature of mechanical drawings, the plan view offers the best illustration of the location and configuration of the work.

Because of the large amount of information required for mechanical work and the close proximity of piping, valves, and connections, mechanical drawings utilize a variety of symbols to convey the intent. These symbols and abbreviations are incorporated in a chart or table called the *legend*. The legend explains symbols and abbreviations shown on the drawings. In addition, the mechanical drawings make use of the schedule format discussed earlier to list such items as the plumbing fixtures, HVAC diffusers, and any appliances.

Electrical Drawings The last part of the working drawings are usually the electrical drawings, which show the various electrical and communication systems of the building. In custom-drawn plans, these sheets are normally supplied by an architect or electrical contractor. In the case of stock plans, the placement of lighting fixtures, switches and outlets

is normally shown, but the power plan and electrical service diagrams are omitted due to differences in local requirements. Again, Home Planners offers generic electrical drawings to illustrate typical components and layouts in a residential electrical system. In all cases, the builder, together with a qualified electrician, should agree upon a specific layout, number of fixtures, and cost prior to completion of the construction estimates.

The electrical drawings include electrical power and lighting plans, telecommunications, and any specialized wiring systems, such as fire or security alarms. The electrical drawings are similar in format to the mechanical drawings in their use of plan view drawings for layout and the use of details and schedules for clarifications. Often power, lighting, and telecommunications layouts are shown on one drawing. For more complex structures, the systems are separated.

The power plan illustrates the power requirements of the structure, locating panels, receptacles, and the circuitry of power-utilizing equipment. Included on the power plan are the panel schedules that list the circuits and power requirements for a project with multiple panels. This is similar to the gas equipment schedule found on the mechanical drawings. Panel schedules list the circuits in the panel, plus the individual power required for each panel and the total power requirement for the electrical system. The schedules total the power

requirements to help the electrical contractor size the panel; then the power company sizes the service requirements. The lighting plan locates the various lighting fixtures in the building and is complemented by the use of a fixture schedule that lists the types of light fixtures to be used. The fixture schedule may list, by number or letter, the manufacturer and model, the wattage of the lamps, voltage, and any special remarks concerning the fixture. The lighting plan locates such devices as switches, smoke and fire detection equipment, emergency lighting, cable TV, and telephone outlets.

Summary

This section is intended as a brief review of the types of plans involved in residential construction. It is essential the builder become familiar with the drawings prior to a detailed site inspection and quantity takeoff. A thorough review of the drawings will reveal important considerations to the spec or custom builder. These include potential construction complexities or high costs of certain features, omissions that must be accounted for in the estimate, or the need for specialized labor or materials.

For more information on plan reading and detailed instructions for material takeoff, refer to Plan Reading and Material Takeoff *by Wayne J. DelPico, 1994, R.S. Means Co., Inc., Kingston, MA 02364.*

Site Selection Overview

The site chosen for a new home is at least as crucial to the sale and profit of the project as the design and quality of workmanship. Selecting a site should be preceded by a thorough market evaluation, which will determine the best location for your market. Factors that affect the location's desirability include price and community (e.g., safety, appearance, proximity to schools and shopping). Features to evaluate for a particular site include privacy, quiet surroundings, a pleasant view, and trees or vegetation that add to its value. Other factors to consider include the price of other homes in the neighborhood and the recent history of home sales in the area.

A careful site visit and examination, together with research of public documents (such as survey reports and topographical maps) will provide other crucial information about a specific site. Is extensive (and expensive) clearing of trees and other vegetation required? Is the land ready to build on? Does it have water and sewer service? If a septic system must be used, has a percolation test been performed, and what were the results? Are utility connections such as gas and electric lines readily available? Are there rock outcroppings or soil contamination? Are there any drainage problems? Are there any restrictions to gaining access to the property? Are there zoning or other restrictions, such as covenants imposed by a previous owner? What is the cost of dealing with shortcomings in any of these areas?

Finances must also be considered in evaluating a piece of property. Factors such as price, availability of funds from a lender for this property, expected turnover time on the sale of the home, and interest on borrowed funds. If the price is significantly different from that of other lots recently sold in the same area, it is important to find out why.

Choosing the right site is fundamental to success in building new homes. The area market analysis, specific site evaluation, and financial analysis and arrangements are each major parts of the site selection process, and builders cannot afford to treat any of them lightly.

Selecting a Spec Building Plan

Once you have done adequate market research on the area in which you intend to build, you can use the information you gathered to make some decisions about the best plan to use. You'll want to choose a plan that fits well with other homes in the neighborhood in terms of size, style, and type. For instance, if the majority of the houses in the area around your site are one-story ranch homes, you should probably avoid a multi-level ultra-contemporary home. Not only would such a style be out of the price range, but you risk alienating area residents, and make it difficult to sell prospective buyers on a house that looks so out of place. You might check into the kinds of houses that other builders in the area have built and sold successfully and adapt your ideas to a similar size, style, and type.

Floor plans are another important consideration. If you are planning to construct a home in a predominantly family neighborhood, you should choose an appropriate floor plan for your spec house. Separate casual and formal living and dining areas and an adequate number of bedrooms are key elements. Conversely, if your market is comprised of empty-nesters, you'll want to look for plans with mostly one-level livability, an all-purpose living area, and space for hobbies or exercise. Starter-home buyers want a simple floor plan that fits their budget and contains options for expandability. Use market research information to determine the most appropriate type of floor plan for your prospective buyers.

It may go without saying, but you'll want to choose plans that are buildable. While today's prospective buyers are looking for amenity-rich plans, they are not always willing to pay for complicated structures or expensive options. Stick with classic designs that stand the test of time. This way, you'll avoid surprises for yourself and your subcontractors, and avoid added expense that buyers may not want to absorb. This does not mean that your spec house has to be overly simplified or boxy in appearance. If you choose your plans from experienced professionals who understand construction as well as good design, you'll find it easy to build a livable home that will attract buyer attention.

The Stock Plan Alternative

One solution to the challenge of finding quality, buildable plans is using stock plans. Since stock plan companies are in the business of creating and selling the most desirable plans, they stay on top of trends through professional associations and independent research. They also have the advantage of garnering daily feedback from their customers about the kinds of plans and features people want most. Stock plan companies are able to respond in an immediate way to a changing market and many can make modifications to designs to meet new trends.

Most stock plan companies have a large, assorted portfolio of designs to choose from. Home Planners, Inc., located in Tucson, Arizona, is based on fifty years of design development and has over 3,000 designs ranging from ultra-modern to historic renditions in every size imaginable. They have access to the portfolios of eight additional designers. It would be very difficult to find a single designer or architect who could offer this same kind of variety.

Working With Stock Plan Companies

Stock plans from an established company such as Home Planners offer the advantages of a wide selection of designs and of easily passing inspection by local building departments, since they have been so carefully reviewed for ease of use. You'll also benefit from the support of a professional staff that can answer questions about the plans.

Home Planners offers auxiliary products and services as well, which can facilitate building projects or enhance buyer interest in homes. A unique customization service can save time and money on modifications to many of the plans. Materials lists and specification outlines help with planning and form the basis for creating legal documents for the construction project. Detailed landscape or deck plans that fit the house plan you've chosen are also available to help put the finishing touches on the home and ensure its success on the market. These services and products are available through Home Planners' wide array of books and through the magazine it publishes and distributes to newsstands across the United States and Canada. The books are readily available at bookstores, supermarkets, libraries, and through mail order. (See the order pages at the back of this book for more information.)

Home Planners has also established a special program for builders—called "The Builder's Edge"—that addresses their unique needs. Included in the program are discounts on books, blueprints and additional materials; design updates; a listing in Home Planners' books and magazines; as well as sales leads.

A word of caution about using stock plan companies: Because blueprints are created in response to your order, they cannot be returned. You must also remember that blueprints cannot be legally copied. You should purchase enough sets to accomplish your building project, or even better, purchase a set of changeable, reproducible sepias.

Procuring the Building Permit

If you have not already procured the building permit forms, you may find it helpful to ask the following questions of the building department as you prepare/assemble the information and documents they require.

- What general information is required on permit forms, and what documents must accompany the forms (*e.g., bank note, deed*)?

- What is the fee schedule for building permits? Is it per real estate value or cost of construction? How is the value determined and by whom? *This is important information needed to predict expenses for your business plan.*

- What is the expected processing time for obtaining a building permit? *This information is key in scheduling the start of work as soon as possible after the permit is issued.*

- Within what period of time must the work be completed to be covered by the building permit? If it is not completed by that date, what is the procedure to obtain an extension? *This may be an additional cost factor in terms of financing and fees.*

- Are plans required and if so, how many copies? What information must be shown on the plans? *The expense and time to produce the required plans must be considered in the project budget and schedule. The benefit of having and submitting detailed plans is not only to show professionalism, but to avoid ambiguities that might cause problems at inspections when the work has already been completed.*

- What building code governs construction in your town? *Knowing this enables you to evaluate plans for code compliance before presenting them to the building department.*

- What is required to obtain the Certificate of Occupancy? *With this information you can be sure all requirements are met so that the owners can move in the day papers are passed.*

- What inspections are required and at what stages of construction? How far in advance should I schedule appointments for inspections? *These are key questions since costs may be increased if work is held up or subcontractors become unavailable as a result of delays.*

While the prior questions are not specific to any particular type of house or site, those that follow begin to address the specific requirements of an individual home.

- Is a registered architect or engineer required to produce a site plan and to locate the structure's boundaries on the lot? *This item must be budgeted and time allowed in the schedule to complete this task.*

- What are the building department's requirements for excavation and the foundation, and is an inspection of the foundation required before backfilling? *This is important information for scheduling the work and budgeting any special construction requirements for the foundation.*

- Are there any miscellaneous testing requirements, such as soil compression or concrete slump? *Again, such tests need to be included in both the schedules and the budget.*

Forms for Home Builders

The following pages are examples of forms that can be useful to builders in organizing and standardizing their projects and businesses. Forms are an important management tool. Consistent use of a set of forms is helpful not only in administering current projects, but for keeping accurate historical records to be used in future estimating and scheduling.

This collection includes a **Letter of Transmittal**, a **Telephone Quotation** form, a **Quantity Sheet** for estimating, a **Proposal** for changes or additional work, an **Extra Work Order**, a **Project Schedule** and a **Job Progress Report** for comparing work completed to the schedule. Each form is preceded by a description of its purpose and instructions for use.

Letter of Transmittal

Use the Letter of Transmittal for many correspondence requirements. The preprinted headings and lists organize information needed for proposal submittals, drawing transmittals, and general communications between contractor, architect, and owner. By filling in the information and checking off the appropriate items, a letter can be completed quickly, without writing and typing an original document.

A blank Letter of Transmittal is shown in Figure 1.

Telephone Quotation

The Telephone Quotation is designed for recording all of the basic information and costs provided by a subcontractor or material vendor for a given job. The form ensures a well organized and consistent record of each bid. It can also function as a basic checklist for the materials and services included in each bid. Using this form makes it easier to compare bids.

Use the one-page form to record all of the necessary bid information – from the basic data such as name, address, telephone number, and contact person – to costs and a description of the work to be completed. Use the checklist part of the form to record any special conditions included in the bid. With this list, delivery charges, sales taxes, and other contingencies and modifications can be clearly delineated and later cross-referenced with other bids. Note that space is also provided to record in detail any of the bidders' exclusions or qualifications to the project's plans and specifications. Note bids for alternates at the end of the form.

A blank Telephone Quotation is shown in Figure 2.

Quantity Sheet

The Quantity Sheet provides an organized method for the quantity takeoff process. First, determine the dimensions for each item from the drawings, and transfer this information to the worksheet. Next, label the three boxes at the head of the "Dimensions" column appropriately. Then, extend the dimensions according to the formula recorded in the boxes at the column head. Totals for quantities of like items are figured by adding down.

A blank Quantity Sheet is shown in Figure 3.

Proposal

The Proposal is used to respond to requests from the buyer for pricing, changes, or additional work to an ongoing project. The form offers an organized, standard document for submitting and accepting a proposal.

Use this form to respond to a request by the buyer for a reduction, addition, or change to the original scope of contracted work. After the form has been completed and signed by both parties, it serves as a mutually agreed upon record of the proposal and its acceptance.

A blank Proposal is shown in Figure 4.

Extra Work Order

The Extra Work Order is used to document orders for extra work. The form, signed by the owner or owner's representative, authorizes the builder to proceed with work not in the original contract. The price can be shown as a unit price or a lump sum. The form helps to ensure that all extras are being completed with the knowledge of the owner, and at a previously agreed upon price. Many times, quick decisions must be made on the job site to expedite the work and avoid delays. This standard form, shown in Figure 5, is good for this purpose.

Project Schedule

The Project Schedule form is designed to help the contractor determine the sequence of activities and the total project duration time. When filled out, the project schedule displays this information in the form of a bar chart.

This form is based on the GANTT, or bar chart, scheduling method. First, fill in days, weeks, or months across the top of the form. Then mark in a bar for each activity listed in the

"Description" column. The length and location of the bar are determined by the activity start date, duration, and finish date. Check the schedule against actual construction progress by adding another bar above the first. The second bar shows the actual activity start date, duration, and completion. Use the bottom two lines of the form for calendar or other information.

While this form, shown in Figure 6, is particularly well suited to developing broad overview schedules, it can also be used for detailed scheduling. It is a useful form for all stages of a project's development — from concept to construction.

Job Progress Report

The Job Progress Report form is set up so that the builder can enter information by work item. For any given job, list work items and their estimated value (in dollars or as a percent). Use the appropriate columns to record the volume of work or percent complete over each period of time. At the bottom of the form is a space for recording the entry date. Use as many sheets as necessary for the duration of the job. This form, shown in Figure 7, is a good tabular record for updating the schedule and developing progress payments.

Means Forms

**LETTER
OF TRANSMITTAL**

FROM

_____

_____

_____

_____

TO

_____

_____

_____

_____

DATE _____

PROJECT _____

LOCATION _____

ATTENTION _____

RE _____

_____

_____

_____

_____

Gentlemen:

WE ARE SENDING YOU ☐ HEREWITH ☐ DELIVERED BY HAND ☐ UNDER SEPARATE COVER

VIA _____ THE FOLLOWING ITEMS:

☐ PLANS ☐ PRINTS ☐ SHOP DRAWINGS ☐ SAMPLES ☐ SPECIFICATIONS

☐ ESTIMATES ☐ COPY OF LETTER ☐ _____

COPIES	DATE OR NO.	DESCRIPTION

THESE ARE TRANSMITTED AS INDICATED BELOW

☐ FOR YOUR USE ☐ APPROVED AS NOTED ☐ RETURN _____ CORRECTED PRINTS

☐ FOR APPROVAL ☐ APPROVED FOR CONSTRUCTION ☐ SUBMIT _____ COPIES FOR _____

☐ AS REQUESTED ☐ RETURNED FOR CORRECTIONS ☐ RESUBMIT _____ COPIES FOR _____

☐ FOR REVIEW AND COMMENT ☐ RETURNED AFTER LOAN TO US ☐ FOR BIDS DUE _____

☐ _____

REMARKS _____

_____

_____

_____

_____

_____

_____

IF ENCLOSURES ARE NOT AS INDICATED,
PLEASE NOTIFY US AT ONCE.

SIGNED _____

Figure 1 Letter of Transmittal

⚓ Means Forms

TELEPHONE QUOTATION

	DATE _____
PROJECT _____	TIME _____
FIRM QUOTING _____	PHONE (____)
ADDRESS _____	BY _____
ITEM QUOTED _____	RECEIVED BY _____

WORK INCLUDED	AMOUNT OF QUOTATION

DELIVERY TIME	**TOTAL BID**	
DOES QUOTATION INCLUDE THE FOLLOWING	If ☐ NO is checked, determine the following	
STATE & LOCAL SALES TAXES ☐ YES ☐ NO	MATERIAL VALUE	
DELIVERY TO THE JOB SITE ☐ YES ☐ NO	WEIGHT	
COMPLETE INSTALLATION ☐ YES ☐ NO	QUANTITY	
COMPLETE SECTION AS PER PLANS & SPECIFICATIONS ☐ YES ☐ NO	DESCRIBE BELOW	
EXCLUSIONS AND QUALIFICATIONS		
ADDENDA ACKNOWLEDGEMENT	**TOTAL ADJUSTMENTS**	
	ADJUSTED TOTAL BID	
ALTERNATES		
ALTERNATE NO.		
ALTERNATE NO.		
ALTERNATE NO.		
ALTERNATE NO.		
ALTERNATE NO.		
ALTERNATE NO.		
ALTERNATE NO.		

Figure 2 Telephone Quotation

Means Forms

QUANTITY SHEET

SHEET NO.

PROJECT

ESTIMATE NO.

LOCATION   ARCHITECT   DATE

TAKEOFF BY   EXTENSIONS BY   CHECKED BY

DESCRIPTION	NO.	DIMENSIONS		UNIT		UNIT		UNIT	

Figure 3 Quantity Sheet

 Means Forms

PROPOSAL

FROM _____

_____
_____
_____
_____

TO

_____
_____
_____
_____

PROPOSAL NO. _____

DATE _____

PROJECT _____

LOCATION _____

_____

CONSTRUCTION TO BEGIN _____

COMPLETION DATE _____

Gentlemen:

The undersigned proposes to furnish all materials and necessary equipment, and perform all labor necessary to complete the following work:

_____
_____
_____
_____
_____
_____
_____
_____
_____
_____
_____
_____
_____
_____

All of the above work to be completed in a substantial and workmanlike manner

☐ for the sum of_____dollars ($_____)

☐ to be paid for at actual cost of Labor, Materials and Equipment plus_____ percent (_____%)

Payments to be made as follows:_____

_____

_____The entire amount of the contract to be paid within_____ after completion.

Any alteration or deviation from the plans and specifications will be executed only upon written orders for same and will be added to or deducted from the sum quoted in this contract. All additional agreements must be in writing.

The Contractor agrees to carry Workers' Compensation and Public Liability Insurance and to pay all taxes on material and labor furnished under this contract as required by Federal laws and the laws of the State in which this work is performed.

Respectfully submitted,

Contractor _____

By_____

ACCEPTANCE

You are hereby authorized to furnish all material, equipment and labor required to complete the work described in the above proposal, for which the undersigned agrees to pay the amount stated in said proposal and according to the terms thereof.

Date _____ 19____

_____
_____

Figure 4 Proposal

⚓ Means Forms

**EXTRA
WORK ORDER**

FROM

_____

_____

_____

_____

TO

_____

_____

_____

_____

EXTRA WORK ORDER NO. _____

DATE _____

PROJECT _____

LOCATION _____

JOB NO. _____

EXTRA WORK ORDER APPROVED

BY _____

BY _____

Gentlemen:

This EXTRA WORK ORDER includes all Material, Labor and Equipment necessary to complete the following work;

☐ the work below to be paid for at actual cost of Labor, Materials and Equipment plus _____ percent (_____%)

☐ the work below to be completed for the sum of_____dollars

($ _____)

DESCRIPTION

_____

_____

_____

_____

_____

_____

_____

_____

_____

_____

_____

_____

_____

_____

_____

_____

_____

_____

_____

The work covered by this order shall be performed under the same Terms and Conditions
as that included in the original contract unless stated otherwise above.

Signed_____

By_____

Figure 5 Extra Work Order

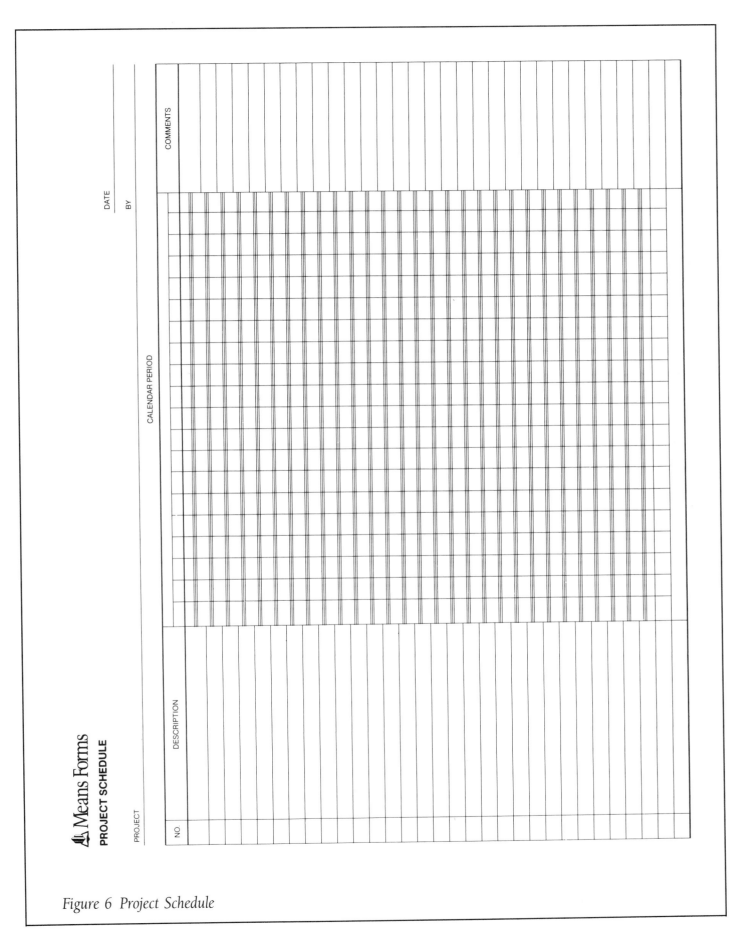

Figure 6 Project Schedule

Means Forms

**JOB
PROGRESS REPORT**

			SHEET	OF

PROJECT   JOB NO.

LOCATION				YEARS

WORK ITEM	QUANTITY, $, OR %	BEGINNING BALANCE	MONTHS	ENDING BALANCE
	DATE			
	YEAR			

Figure 7 Job Progress Report

Safety Tips

This section is included because safety is a crucial element in any construction project. Proper and thorough measures are often overlooked as a result of time pressures and carelessness. The following list is a review/reminder of some basic rules that should be applied to every project.

- Make a point of keeping the work area neat and organized. Eliminate tripping hazards and clutter, especially in the areas used for access. Take the time to clean up and reorganize as you go along. This means not only keeping neat piles of materials, but also remembering not to leave nails sticking out; pull them or bend them over. Remember, you may be called away or run into difficulties that delay the completion of the project, and should not leave a hazard. Also, potential buyers could arrive at any time.

- Wear the proper clothing and gear to protect yourself from possible hazards.

- Avoid loose or torn clothing if you are working with power tools. Serious injury can be caused if clothing is caught in moving parts.

- Wear heavy shoes or boots to protect your feet, and a hard hat in any situation where material or tools could fall on your head.

- Wear safety glasses when working with power tools or in any other circumstance where there is the potential for injury to the eyes.

- Use hearing protection when operating loud machinery or when hammering in a small, enclosed space.

- Wear a dust mask to protect yourself from inhaling sawdust, insulation fibers, or other airborne particles.

- Wear suitable gloves whenever possible to minimize hand injuries.

- When lifting, always try to let your leg muscles do the work, not your back. Keep your back straight, your chin tucked in, and your stomach pulled in. Maintain the same posture when setting an item down. Seek assistance when moving heavy or awkward objects and, remember, if an object is on wheels, it is easier to push than to pull it.

- When working from a ladder, scaffold, or any temporary platform, make sure it is stable and well braced. When walking on joists, trusses, or rafters, always watch each step to see that what you are stepping on is secure.

- Take the extra time to install barriers around floor openings.

- When working with adhesives, protective coatings, or other volatile (fume-emitting) products, be sure to follow manufacturers' recommendations on proper ventilation. Pay particular attention to drying times and fire hazards associated with the product. If possible, obtain from your supplier a Material Safety Data Sheet, which will clearly describe any associated hazards.

- Do not use or store flammable liquids or use gasoline-fueled tools or equipment inside of a building or enclosed area.

• When working with electricity or gas, be sure you know how to shut off the supply; then make sure it is off in a safe way by testing the outlet fixture or equipment. It may be wise to invest in a simple current-testing device to determine when electric current is present. Be sure you know how to use it properly. If you don't already have one, purchase a fire extinguisher, learn how to use it, and keep it handy. Have emergency telephone numbers and utility telephone numbers at hand.

• When using power tools, never pin back safety guards. Choose the correct cutting blade for the material you are using. Keep children or bystanders away from the work area, and never interrupt someone using a power tool or actively performing an operation. Always unplug tools when leaving them unattended or when servicing or changing blades.

A few tips on hand tools:

• Do not use any tool for a purpose other than the one for which it was designed. In other words, do not use a screwdriver as a pry bar, pliers as a hammer, etc. Not only can the tools be easily ruined, but the impact that causes the damage may also injure the user.

• Do not use any striking tool (such as a hammer or sledgehammer) that has dents, cracks, or chips; shows excessive wear; or has a damaged or loose handle. Also, do not strike a hammer with another hammer in an attempt to remove a stubborn nail, get at an awkward spot, etc. Do not strike hard objects like concrete or steel, which could chip the hammer, causing personal injury.

• If you rent or borrow tools and equipment, take time to read the instructions or obtain proper instruction from an experienced person.

Seek further advice on proper tool selection and use from your local building supply dealer, or from the Hand Tools Institute at 25 N. Broadway, Tarrytown, NY 10591. Always review manufacturers' instructions and warnings.

Section 2
Plans with Estimates

How to Use Plans & Estimates

The following is a detailed explanation of a sample plan and cost estimate. The two-page spread includes an illustration of the house and a floor plan along with the components and related costs of that plan.

Description
Information about each plan is listed here concerning the total living area, number of bedrooms and bathrooms, and the price schedule of blueprints.

Plan Information
Each plan has been assigned a unique identification number. Also included here is the style and number of stories of the plan.

Illustration
Each plan includes a rendering of the house costed on the facing page. Please note that landscaping is not included in the cost.

Floor Plan
This is a floor plan view of the house costed on the facing page.

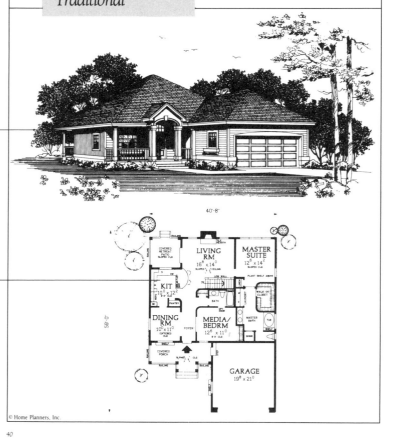

RM-3442
1 Story
Traditional

1273 Square Feet
2 Bedrooms
2 Baths
Schedule A

© Home Planners, Inc.

40

Contractor's Overhead and Profit

These figures include overhead costs for the general contractor and profit on materials, labor and equipment costs.

Components

This page contains the ten categories of components needed to develop the complete square foot cost of the plan shown. All components are defined with a description of the materials and/or task involved. Use cost figures from each category to estimate the cost per square foot for that section of the project.

RM-3442
Cost Estimate

Cost at a Glance

Cost per Square Foot: $120.51
Total Cost: $153,409

Cost by Category		Cost per Square Foot of Living Area		
		Materials	Installation	Total
1. Site Work	Excavation for the basement and footings.		1.32	1.32
2. Foundation	Main house – 6″ and 10″ wide reinforced concrete foundation wall on 16″ x 10″ and 20″ x 10″ reinforced concrete perimeter footings. Trench footings – 8″ wide reinforced concrete. Slabs – 4″ thick reinforced steel trowel finished concrete over compacted gravel.	5.76	7.06	12.82
3. Framing	Exterior walls – 2 x 6 studs, 16″ on center with 1/2″ plywood sheathing. Floors – 2 x 10 floor joists, 16″ on center with 3/4″ plywood subfloor. Roof – site cut 2 x 12 rafters and pre-engineered trusses with 5/8″ sheathing.	19.43	17.07	36.50
4. Exterior Walls	Beveled cedar siding and Texture 1-11 over 15# felt vapor barrier with R-19 and R-11 wall insulation. Vinyl clad fixed, and double hung windows and swinging patio doors.	13.37	4.35	17.72
5. Roofing	Heavyweight three tab asphalt shingles over 30# felt roofing paper. Aluminum drip edge and flashings.	2.37	2.11	4.48
6. Interiors	Walls and ceilings – 1/2″ and 5/8″ taped and finished gypsum wallboard, primed and painted with one coat latex. Pine interior trim, with one coat paint or stain. Flooring – 70% carpet, 8% vinyl, 7% hardwood, and 15% ceramic tile.	7.93	7.64	15.57
7. Specialties	Hardwood faced particle board case kitchen cabinets and bathroom vanities with plastic laminate countertops. Washer, dryer, range with hood, dishwasher, and refrigerator. One pre-fabricated fireplace.	7.54	2.02	9.56
8. Mechanical	Oil fired forced hot air heat with central air conditioning. One full bath and a master suite with a tub and shower. Stainless steel double bowl kitchen sink with disposal.	6.17	3.66	9.83
9. Electrical	200 amp service, branch circuit wiring with romex cable. Exterior and interior lighting fixtures, receptacles and switches.	3.03	1.80	4.83
10. Overhead	Contractor's overhead and profit.	4.59	3.29	7.88
Total Cost per Square Foot		$70.19	$50.32	$120.51

To purchase a full set of Sepias, Bill of Materials and Detailed Costs – turn to page 267.

41

Cost at a Glance

These figures give quick overall values for this plan.

Materials

This column contains the Materials Cost of each component.

Installation

Installation includes labor and equipment plus the installing contractor's overhead and profit. The average labor mark-up used to create these figures is 50.95% over and above **Bare Labor Cost** including fringe benefits.

Total

The figure in this column is the sum of the material and installation costs.

Bottom Line Total

This figure is the complete square foot cost for the plan shown. To determine **Total Project Cost**, multiply the **Bottom Line Total** times the **Living Area**. (Total Project Cost = Botttom Line Total × Living Area)

Note: See Location Factors at the end of this section to adjust costs to a specific location.

Up to 1500 Square Feet

Plans	Style	Stories	Total SF	Bedrms	Baths	Page
RM2707	Colonial	1 Story	1267	3	2	28
RM2864	Contemporary	1 Story	1387	3	2	30
RM3460B	Farmhouse	1 Story	1389	3	2	32
RM3373	Ranch	1 Story	1378	3	2	34
RM3355	Ranch	1 Story	1387	3	2	36
RM3375	Southwestern	1 Story	1378	3	2	38
RM3442	Traditional	1 Story	1273	2	2	40
RM2505A	Traditional	1 Story	1366	3	2	42
RM3374	Tudor	1 Story	1378	3	2	44
RM2606A	Tudor	1 Story	1499	3	2½	46
RM2622	Colonial	2 Story	1248	3	2½	48

RM-2707

1 Story Colonial

1·267 Square Feet
3 Bedrooms
2 Baths
Schedule A

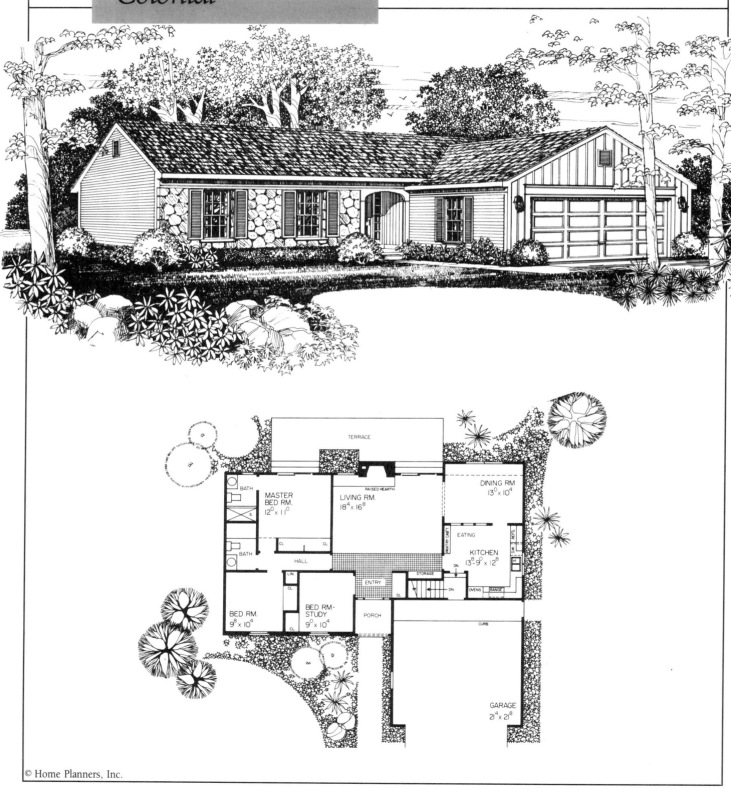

RM-2707
Cost Estimate

Cost at a Glance

Cost per Square Foot: $91.10
Total Cost: $115,423

Cost by Category

Category	Description	Cost per Square Foot of Living Area		
		Materials	Installation	Total
1. Site Work	Excavation for the basement and footings.		1.32	1.32
2. Foundation	Main house — 12″ wide concrete block foundation wall on 20″ x 10″ reinforced concrete perimeter footings. Trench footings — 8″ wide reinforced concrete. Slabs — 4″ thick reinforced steel trowel finished concrete over compacted gravel.	5.00	6.65	11.65
3. Framing	Exterior walls — 2 x 6 studs, 16″ on center with 1/2″ plywood sheathing. Garage — 2 x 4 studs, 16″ on center. Floors — 2 x 8 floor joists, 16″ on center with 3/4″ plywood subfloor. Roof — pre-engineered trusses and site cut rafters with 5/8″ plywood sheathing.	8.75	5.53	14.28
4. Exterior Walls	Beveled cedar siding and vertical pine boards over 15# felt vapor barrier with R-19 and R-11 wall insulation. Vinyl clad casement and double hung windows and sliding patio doors.	12.71	4.97	17.68
5. Roofing	Heavyweight three tab asphalt shingles over 30# felt roofing paper. Aluminum gutters, downspouts, drip edge and flashings.	2.15	1.65	3.80
6. Interiors	Walls and ceilings — 1/2″ and 5/8″ taped and finished gypsum wallboard, primed and painted with one coat latex. Pine interior trim. Flooring — 78% carpet, 7% hardwood, 10% vinyl, and 5% ceramic tile.	7.88	6.55	14.43
7. Specialties	Hardwood faced particle board case kitchen cabinets and bathroom vanities with plastic laminate countertops. Washer, dryer, cooktop with hood, double ovens, dishwasher, and refrigerator. One masonry fireplace.	6.84	2.76	9.60
8. Mechanical	Oil fired forced hot air heat with central air conditioning. One full bath and one 3/4 bath. Stainless steel double bowl kitchen sink with disposal.	4.76	3.26	8.02
9. Electrical	200 amp service, branch circuit wiring with romex cable. Exterior and interior lighting fixtures, receptacles and switches.	2.78	1.58	4.36
10. Overhead	Contractor's overhead and profit.	3.56	2.40	5.96
Total Cost per Square Foot		$54.43	$36.67	$91.10

To purchase a full set of Sepias, Bill of Materials and Detailed Costs — turn to page 267.

RM-2864
1 Story Contemporary

1387 Square Feet
3 Bedrooms
2 Baths
Schedule A

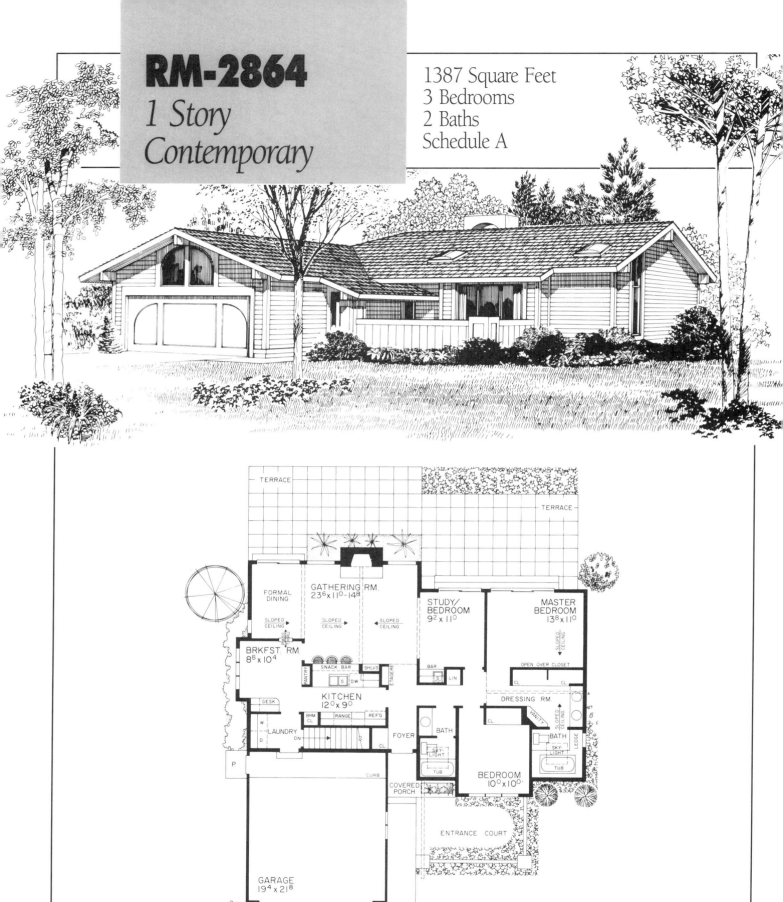

TERRACE

TERRACE

FORMAL DINING

GATHERING RM.
23^6 x 11^0-14^8

STUDY/ BEDROOM
9^2 x 11^0

MASTER BEDROOM
13^8 x 11^0

SLOPED CEILING

SLOPED CEILING

SLOPED CEILING

SLOPED CEILING

BRKFST. RM.
8^8 x 10^4

SNACK BAR

SHLVS

ETAGERE

PANTRY

BAR

LIN

OPEN OVER CLOSET

CL

CL

DRESSING RM.

DESK

KITCHEN
12^0 x 9^0

BRM CL

RANGE

REF'G

FOYER

BATH

CL

VANITY

SLOPED CEILING

BATH

LEDGE

W

LAUNDRY

D

DN

SKY- LIGHT

TUB

SKY- LIGHT

TUB

P

CURB

BEDROOM
10^0 x 10^0

COVERED PORCH

ENTRANCE COURT

GARAGE
19^4 x 21^8

RM-2864
Cost Estimate

Cost by Category

Category	Description	Cost per Square Foot of Living Area		
		Materials	Installation	Total
1. Site Work	Excavation for the basement and footings.		1.27	1.27
2. Foundation	Main house — 12″ concrete block wall on a 20″ x 10″ reinforced concrete footing. Garage — 8″ x 42″ reinforced concrete foundation wall. Slabs — 4″ thick reinforced steel trowel finished concrete on 4″ compacted gravel.	5.13	6.71	11.84
3. Framing	Main house — 2 x 6 studs, 16″ on center with 1/2″ plywood sheathing. Garage — 2 x 4 studs, 16″ on center. Floor — 2 x 10 joists, 16″ on center with 3/4″ tongue and groove plywood subfloor. Roof — site cut 2 x 12 rafters with 1/2″ plywood sheathing.	14.79	8.46	23.25
4. Exterior Walls	Bevel cedar siding and Texture 1-11 with 15# felt vapor barrier. Vinyl clad fixed and casement windows and sliding glass doors. Flush entry doors. R-19 and R-11 insulation.	15.83	6.59	22.42
5. Roofing	Heavyweight three tab asphalt roof shingles on 30# felt paper. Two dome type skylights. Aluminum drip edge, flashings, gutters and downspouts.	2.84	2.14	4.98
6. Interiors	Wall finish — one coat primer and one coat paint on 1/2″ or 5/8″ gypsum wallboard. Pine door, window and baseboard moldings. Flooring — 70% carpet, 19% vinyl, 4% ceramic tile and 7% hardwood.	7.34	6.32	13.66
7. Specialties	Hardwood faced, particle board case kitchen cabinets and bath vanities with plastic laminate countertops. Washer, dryer, range, dishwasher and refrigerator. Pre-fabricated fireplace, a covered porch and terrace.	7.89	2.32	10.21
8. Mechanical	Oil fired forced hot air heat with central air conditioning. Two full baths. Double bowl kitchen sink with disposal.	5.33	3.21	8.54
9. Electrical	200 amp service, branch circuit wiring with romex cable. Exterior and interior lighting fixtures, receptacles and switches.	2.89	1.53	4.42
10. Overhead	Contractor's overhead and profit.	4.34	2.70	7.04
	Total Cost per Square Foot	$66.38	$41.25	$107.63

To purchase a full set of Sepias, Bill of Materials and Detailed Costs — turn to page 267.

RM-3460B

1 Story Farmhouse

1389 Square Feet
3 Bedrooms
2 Baths
Schedule A

FAMILY RM
VAULTED CLG
12^4 x 12^0

MASTER BEDRM
VAULTED CLG
13^0 x 12^0

MASTER BATH

BEDRM
VAULTED CLG
10^0 x 10^8

SNACK BAR

PANTRY

D   W

LAUNDRY

KIT
VAULTED CLG
12^4 x 10^0

SINK

REFG

LINEN

BEDRM
VAULTED CLG
10^0 x 10^8

COVERED PORCH

DINING

BATH

LIVING RM
VAULTED CLG
13^{10} x 19^0

PLANT SHELF ABOVE

F/A/U   W.H.

CURB

ENTRY

HALF WALL

COVERED PORCH

GARAGE
21^4 x 23^8

44'-8"

54'-6"

RM-3460B
Cost Estimate

Cost at a Glance

Cost per Square Foot: $76.70
Total Cost: $106,536

Cost by Category

		Cost per Square Foot of Living Area		
		Materials	Installation	Total
1. Site Work	Excavation for the slab and footings.		.48	.48
2. Foundation	Main house—6″ x 24″ reinforced concrete foundation wall on 16″ x 10″ reinforced concrete footings. Slabs—4″ thick reinforced steel trowel finished concrete on 4″ compacted gravel.	2.58	3.37	5.95
3. Framing	Main house and garage—2 x 6 studs, 16″ on center with 1/2″ plywood sheathing. Roof—2 x 8 site cut rafters and pre-engineered trusses with 5/8″ plywood sheathing.	7.37	5.18	12.55
4. Exterior Walls	Vertical board and exterior plywood siding. Vinyl clad single hung and sliding windows and sliding glass doors. Paneled and flush entry doors. R-19 and R-11 insulation.	11.35	3.84	15.19
5. Roofing	Heavyweight three tab asphalt shingles over 30# felt building paper. Aluminum drip and step flashings, gutters and downspouts.	2.39	1.73	4.12
6. Interiors	Wall finish—1/2″ or 5/8″ gypsum wallboard with one coat primer and one coat finish paint. Pine door, window and baseboard moldings, with one coat paint or stain. Flooring—77% carpet, 33% vinyl.	6.86	6.24	13.10
7. Specialties	Hardwood faced, particle board case kitchen cabinets and bath vanities with plastic laminate countertops. Washer, dryer, range with hood, double ovens, dishwasher and refrigerator. Pre-fabricated fireplace.	6.69	1.61	8.30
8. Mechanical	Oil fired forced hot air heat with central air conditioning. Two full baths. Double bowl kitchen sink with disposal.	4.60	3.04	7.64
9. Electrical	200 amp service, branch circuit wiring with romex cable. Exterior and interior lighting fixtures, receptacles and switches.	2.82	1.53	4.35
10. Overhead	Contractor's overhead and profit.	3.13	1.89	5.02
Total Cost per Square Foot		$47.79	$28.91	$76.70

To purchase a full set of Sepias, Bill of Materials and Detailed Costs—turn to page 267.

RM-3373

1 Story Ranch

1378 Square Feet
3 Bedrooms
2 Baths
Schedule A

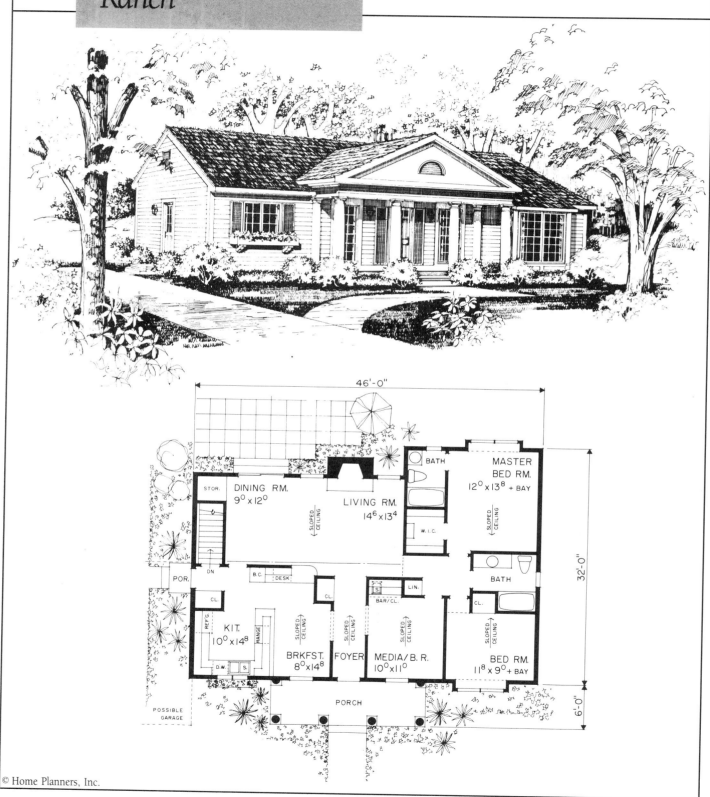

46'-0"

32'-0"

6'-0"

STOR.

DINING RM.
9⁰ x 12⁰

LIVING RM.
14⁶ x 13⁴

SLOPED CEILING

BATH

MASTER BED RM.
12⁰ x 13⁸ + BAY

SLOPED CEILING

W.I.C.

DN

B.C. DESK

CL.

BATH

POR.

CL.

S.

LIN.

BAR/CL.

CL.

REF'G

RANGE

KIT.
10⁰ x 14⁸

SLOPED CEILING

SLOPED CEILING

SLOPED CEILING

SLOPED CEILING

D.W. S.

BRKFST.
8⁰ x 14⁸

FOYER

MEDIA/B. R.
10⁰ x 11⁰

BED RM.
11⁸ x 9⁰ + BAY

PORCH

POSSIBLE GARAGE

RM-3373
Cost Estimate

Cost per Square Foot: *$90.01*
Total Cost: *$124,033*

Cost by Category

		Cost per Square Foot of Living Area		
		Materials	Installation	Total
1. Site Work	Excavation for the basement and footings.		.95	.95
2. Foundation	Main house — 10″ wide reinforced concrete foundation wall on 20″ x 10″ reinforced concrete perimeter footings. Trench footings — 10″ wide reinforced concrete walls. Slabs — 4″ thick steel trowel finished reinforced concrete over compacted gravel.	4.70	6.23	10.93
3. Framing	Exterior walls — 2 x 6 studs, 16″ on center with 1/2″ plywood sheathing. Floor — 2 x 10 joists, 16″ on center with 3/4″ plywood subfloor. Roof — pre-engineered trusses and site cut rafters with 1/2″ plywood sheathing.	11.96	5.35	17.31
4. Exterior Walls	Beveled cedar siding over 15# felt vapor barrier with R-19 insulation. Vinyl clad fixed, double hung and casement windows, and a sliding glass patio door.	10.51	4.63	15.14
5. Roofing	Heavyweight three tab asphalt shingles over 30# felt roofing paper. Aluminum gutters, downspouts, drip edge and flashings.	1.56	1.35	2.91
6. Interiors	Walls and ceilings — 1/2″ and 5/8″ taped and finished gypsum wallboard, primed and painted with one coat latex. Pine interior trim with one coat paint or stain. Flooring — 68% carpet, 23% vinyl, 5% hardwood and 4% ceramic tile.	6.44	5.76	12.20
7. Specialties	Hardwood faced particle board case kitchen cabinets and bathroom vanities with plastic laminate countertops. Washer, dryer, range with hood, dishwasher and refrigerator. One masonry fireplace.	10.03	2.81	12.84
8. Mechanical	Oil fired forced hot air heat with central air conditioning. Two full baths. Stainless steel kitchen sink with disposal.	4.81	3.08	7.89
9. Electrical	200 amp service, branch circuit wiring with romex cable. Exterior and interior lighting fixtures, receptacles and switches.	2.52	1.43	3.95
10. Overhead	Contractor's overhead and profit.	3.68	2.21	5.89
Total Cost per Square Foot		$56.21	$33.80	$90.01

To purchase a full set of Sepias, Bill of Materials and Detailed Costs — turn to page 267.

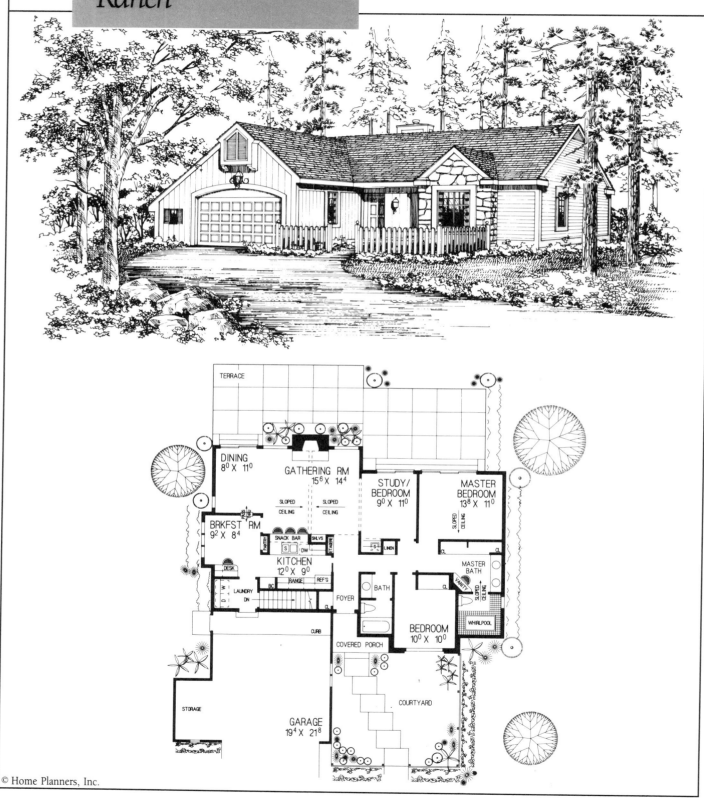

RM-3355

*1 Story
Ranch*

1387 Square Feet
3 Bedrooms
2 Baths
Schedule A

TERRACE

DINING
8⁰ X 11⁰

GATHERING RM
15⁶ X 14⁴

SLOPED CEILING   SLOPED CEILING

STUDY/
BEDROOM
9⁰ X 11⁰

MASTER
BEDROOM
13⁸ X 11⁰

SLOPED CEILING

BRKFST RM
9² X 8⁴

PASS THRU

PANTRY   SNACK BAR   SHLVS
S   DW   STORAGE

LINEN

CL

MASTER
BATH

VANITY

DESK

KITCHEN
12⁰ X 9⁰

RANGE   REF'G

W   D   BC   LAUNDRY   DN

BATH

CL   SLOPED CEILING

FOYER

CL

WHIRLPOOL

BEDROOM
10⁰ X 10⁰

CURB

COVERED PORCH

STORAGE

COURTYARD

GARAGE
19⁴ X 21⁸

RM-3355
Cost Estimate

Cost at a Glance

Cost per Square Foot: $101.30
Total Cost: $140,503

Cost by Category

| | | Cost per Square Foot of Living Area | | |
		Materials	Installation	Total
1. Site Work	Excavation for the basement and footings.		1.27	1.27
2. Foundation	Main house—10″ wide reinforced concrete foundation wall on 20″ x 10″ reinforced concrete perimeter footings. Garage walls—8″ wide reinforced concrete. Slabs—4″ thick steel trowel finished reinforced concrete over compacted gravel.	5.97	7.71	13.68
3. Framing	Exterior walls—2 x 6 studs, 16″ on center with 1/2″ plywood sheathing. Garage—2 x 4 studs, 16″ on center. Floor—2 x 10 joists, 16″ on center with 3/4″ plywood subfloor. Roof— pre-engineered trusses and site cut rafters with 1/2″ plywood sheathing.	12.34	6.64	18.98
4. Exterior Walls	Texture 1-11 and beveled cedar siding over 15# felt vapor barrier with R-19 and R-11 insulation. Vinyl clad fixed, double hung and casement windows, and sliding glass patio doors.	15.24	4.09	19.33
5. Roofing	Heavyweight three tab asphalt shingles over 30# felt roofing paper. Aluminum gutters, downspouts, drip edge and flashings.	2.60	1.81	4.41
6. Interiors	Walls and ceilings—1/2″ and 5/8″ taped and finished gypsum wallboard, primed and painted with one coat latex. Pine interior trim. Flooring—63% carpet, 16% vinyl, 14% hardwood and 7% ceramic tile.	7.28	6.75	14.03
7. Specialties	Hardwood faced particle board case kitchen cabinets and bathroom vanities with plastic laminate countertops. Washer, dryer, range with hood, dishwasher and refrigerator. One pre-fabricated fireplace.	7.53	2.30	9.83
8. Mechanical	Oil fired forced hot air heat with central air conditioning. Two full baths. Stainless steel kitchen sink with disposal.	5.51	3.22	8.73
9. Electrical	200 amp service, branch circuit wiring with romex cable. Exterior and interior lighting fixtures, receptacles and switches.	2.85	1.57	4.42
10. Overhead	Contractor's overhead and profit.	4.15	2.47	6.62
	Total Cost per Square Foot	**$63.47**	**$37.83**	**$101.30**

To purchase a full set of Sepias, Bill of Materials and Detailed Costs—turn to page 267.

RM-3375

1 Story
Southwestern

1378 Square Feet
3 Bedrooms
2 Baths
Schedule A

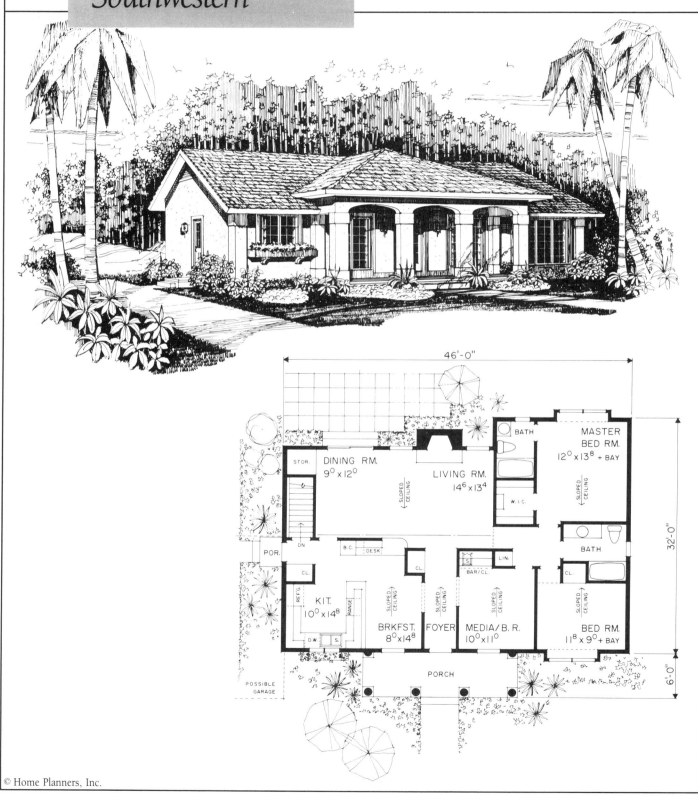

RM-3375
Cost Estimate

Cost at a Glance

Cost per Square Foot: $80.79
Total Cost: $111,328

Cost by Category

		Cost per Square Foot of Living Area		
		Materials	Installation	Total
1. Site Work	Excavation for the basement and footings.		.16	.16
2. Foundation	Main house—10″ wide reinforced concrete foundation wall on 20″ x 10″ reinforced concrete perimeter footings. Trench footings—10″ wide reinforced concrete. Slabs—4″ thick reinforced steel trowel finished concrete over compacted gravel.	4.71	6.02	10.73
3. Framing	Exterior walls—2 x 6 studs, 16″ on center with 1/2″ plywood sheathing. Floor—2 x 10 floor joists, 16″ on center with 3/4″ plywood subfloor. Roof—pre-engineered trusses and site cut rafters with 1/2″ plywood sheathing.	8.67	5.66	14.33
4. Exterior Walls	1″ stucco siding on 3/8″ high rib metal lath over 30# felt vapor barrier with R-19 wall insulation. Vinyl clad fixed and casement windows and a sliding glass patio door.	8.23	3.08	11.31
5. Roofing	Heavyweight three tab asphalt shingles over 30# felt roofing paper. Aluminum gutters, downspouts, drip edge and flashings.	1.39	1.36	2.75
6. Interiors	Walls and ceilings—1/2″ and 5/8″ taped and finished gypsum wallboard, primed and painted with one coat latex. Pine interior trim, with one coat paint or stain. Flooring—68% carpet, 23% vinyl, 5% hardwood and 4% ceramic tile.	6.56	6.08	12.64
7. Specialties	Hardwood faced particle board case kitchen cabinets and bathroom vanities with plastic laminate countertops. Washer, dryer, range with hood, dishwasher, and refrigerator. One masonry fireplace.	9.15	2.55	11.70
8. Mechanical	Oil fired forced hot air heat with central air conditioning. Two full baths. Stainless steel double bowl kitchen sink with disposal.	4.78	3.07	7.85
9. Electrical	200 amp service, branch circuit wiring with romex cable. Exterior and interior lighting fixtures, receptacles and switches.	2.59	1.44	4.03
10. Overhead	Contractor's overhead and profit.	3.23	2.06	5.29
Total Cost per Square Foot		$49.31	$31.48	$80.79

To purchase a full set of Sepias, Bill of Materials and Detailed Costs—turn to page 267.

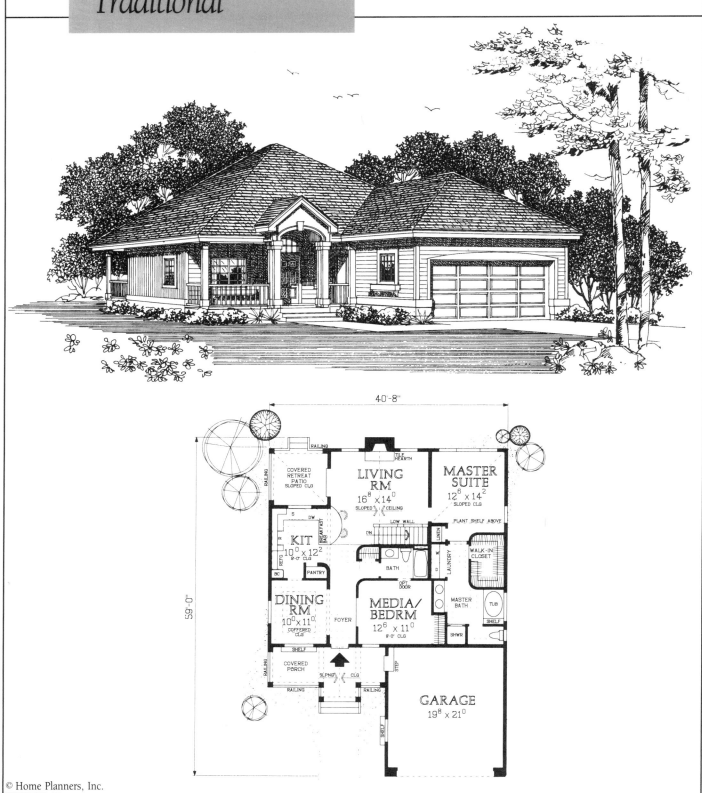

RM-3442

1 Story
Traditional

1273 Square Feet
2 Bedrooms
2 Baths
Schedule A

40'-8"

59'-0"

RAILING

COVERED
RETREAT
PATIO
SLOPED CLG

RAILING

LIVING
RM
16⁸ x 14⁰
SLOPED ✕ CEILING

TILE
HEARTH

MASTER
SUITE
12⁶ x 14²
SLOPED CLG

S   DW

R

KIT
10⁰ x 12²
9'-0" CLG

REFG

BC

PANTRY

BREAKFAST
BAR

LOW WALL

LINEN

PLANT SHELF ABOVE

WALK-IN
CLOSET

BATH

W

D

LAUNDRY

OPT.
DOOR

DINING
RM
10⁰ x 11⁰
COFFERED
CLG

FOYER

MEDIA/
BEDRM
12⁶ x 11⁰
9'-0" CLG

MASTER
BATH

TUB

SHELF

SHWR

SHELF

COVERED
PORCH

RAILING

SLPNG   CLG

RAILING

STEP

RAILING

GARAGE
19⁸ x 21⁰

SHELF

RM-3442
Cost Estimate

Cost at a Glance

Cost per Square Foot: $120.51
Total Cost: $153,409

Cost by Category

		Cost per Square Foot of Living Area		
		Materials	Installation	Total
1. Site Work	Excavation for the basement and footings.		1.32	1.32
2. Foundation	Main house — 6″ and 10″ wide reinforced concrete foundation wall on 16″ x 10″ and 20″ x 10″ reinforced concrete perimeter footings. Trench footings — 8″ wide reinforced concrete. Slabs — 4″ thick reinforced steel trowel finished concrete over compacted gravel.	5.76	7.06	12.82
3. Framing	Exterior walls — 2 x 6 studs, 16″ on center with 1/2″ plywood sheathing. Floors — 2 x 10 floor joists, 16″ on center with 3/4″ plywood subfloor. Roof — site cut 2 x 12 rafters and pre-engineered trusses with 5/8″ sheathing.	19.43	17.07	36.50
4. Exterior Walls	Beveled cedar siding and Texture 1-11 over 15# felt vapor barrier with R-19 and R-11 wall insulation. Vinyl clad fixed, and double hung windows and swinging patio doors.	13.37	4.35	17.72
5. Roofing	Heavyweight three tab asphalt shingles over 30# felt roofing paper. Aluminum drip edge and flashings.	2.37	2.11	4.48
6. Interiors	Walls and ceilings — 1/2″ and 5/8″ taped and finished gypsum wallboard, primed and painted with one coat latex. Pine interior trim, with one coat paint or stain. Flooring — 70% carpet, 8% vinyl, 7% hardwood, and 15% ceramic tile.	7.93	7.64	15.57
7. Specialties	Hardwood faced particle board case kitchen cabinets and bathroom vanities with plastic laminate countertops. Washer, dryer, range with hood, dishwasher, and refrigerator. One pre-fabricated fireplace.	7.54	2.02	9.56
8. Mechanical	Oil fired forced hot air heat with central air conditioning. One full bath and a master suite with a tub and shower. Stainless steel double bowl kitchen sink with disposal.	6.17	3.66	9.83
9. Electrical	200 amp service, branch circuit wiring with romex cable. Exterior and interior lighting fixtures, receptacles and switches.	3.03	1.80	4.83
10. Overhead	Contractor's overhead and profit.	4.59	3.29	7.88
Total Cost per Square Foot		$70.19	$50.32	$120.51

To purchase a full set of Sepias, Bill of Materials and Detailed Costs — turn to page 267.

RM-2505A

1 Story
Traditional

1366 Square Feet
3 Bedrooms
2 Baths
Schedule A

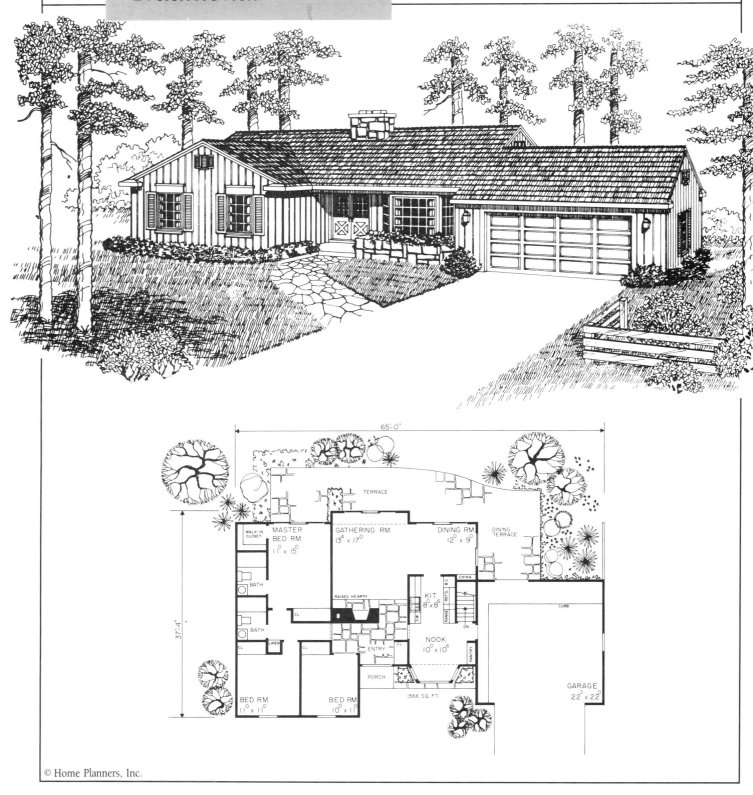

RM-2505A
Cost Estimate

Cost at a Glance

Cost per Square Foot: $94.65
Total Cost: $129,291

Cost by Category

		Cost per Square Foot of Living Area		
		Materials	Installation	Total
1. Site Work	Excavation for the basement and footings.		1.27	1.27
2. Foundation	Main house — 12″ wide concrete masonry unit foundation wall on 20″ x 10″ reinforced concrete perimeter footings. Trench footings — 8″ wide reinforced concrete. Slabs — 4″ thick steel trowel finished reinforced concrete over compacted gravel.	5.84	7.74	13.58
3. Framing	Exterior walls — 2 x 4 studs, 16″ on center with 1/2″ plywood sheathing. Floor — 2 x 10 joists, 16″ on center with 3/4″ plywood subfloor. Roof — pre-engineered trusses and site cut rafters with 1/2″ plywood sheathing.	10.26	6.88	17.14
4. Exterior Walls	Vertical board siding over 15# felt vapor barrier with R-19 and R-11 insulation. Vinyl clad fixed and double hung windows and sliding glass patio doors.	15.63	4.82	20.45
5. Roofing	Heavyweight three tab asphalt shingles over 30# felt roofing paper. Aluminum gutters, downspouts, drip edge and flashings.	1.92	1.74	3.66
6. Interiors	Walls and ceilings — 1/2″ and 5/8″ taped and finished gypsum wallboard, primed and painted with one coat latex. Pine interior trim. Flooring — 75% carpet, 15% vinyl, and 10% stone or ceramic tile.	6.95	5.75	12.70
7. Specialties	Hardwood faced particle board case kitchen cabinets and bathroom vanities with plastic laminate countertops. Washer, dryer, range with hood, dishwasher and refrigerator. One masonry fireplace.	5.57	2.48	8.05
8. Mechanical	Oil fired forced hot air heat with central air conditioning. Two full baths. Stainless steel kitchen sink with disposal.	4.54	3.05	7.59
9. Electrical	200 amp service, branch circuit wiring with romex cable. Exterior and interior lighting fixtures, receptacles and switches.	2.64	1.38	4.02
10. Overhead	Contractor's overhead and profit.	3.73	2.46	6.19
Total Cost per Square Foot		$57.08	$37.57	$94.65

To purchase a full set of Sepias, Bill of Materials and Detailed Costs — turn to page 267.

RM-3374

1 Story
Tudor

1378 Square Feet
3 Bedrooms
2 Baths
Schedule A

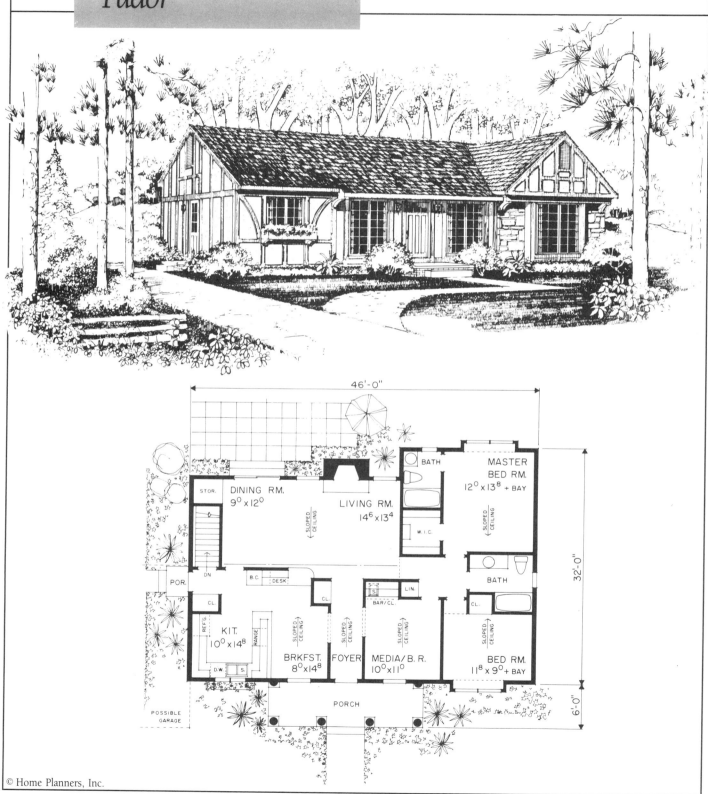

RM-3374
Cost Estimate

Cost at a Glance

Cost per Square Foot: $82.31
Total Cost: $113,423

Cost by Category

		Cost per Square Foot of Living Area		
		Materials	Installation	Total
1. Site Work	Excavation for the basement and footings.		.95	.95
2. Foundation	Main house — 10″ wide reinforced concrete foundation wall on 20″ x 10″ reinforced concrete perimeter footings. Trench footings — 10″ wide reinforced concrete walls. Slabs — 4″ thick steel trowel finished reinforced concrete over compacted gravel.	4.82	6.37	11.19
3. Framing	Exterior walls — 2 x 6 studs, 16″ on center with 1/2″ plywood sheathing. Floor — 2 x 10 joists, 16″ on center with 3/4″ plywood subfloor. Roof — pre-engineered trusses and site cut rafters with 1/2″ plywood sheathing.	7.84	4.88	12.72
4. Exterior Walls	Stucco and cultured stone siding with R-19 insulation. Vinyl clad fixed and casement windows, and a sliding glass patio door.	9.00	4.10	13.10
5. Roofing	Heavyweight three tab asphalt shingles over 30# felt roofing paper. Aluminum gutters, downspouts, drip edge and flashings.	1.27	1.20	2.47
6. Interiors	Walls and ceilings — 1/2″ and 5/8″ taped and finished gypsum wallboard, primed and painted with one coat latex. Pine interior trim with one coat paint or stain. Flooring — 68% carpet, 23% vinyl, and 9% ceramic tile.	6.44	5.74	12.18
7. Specialties	Hardwood faced particle board case kitchen cabinets and bathroom vanities with plastic laminate countertops. Washer, dryer, range with hood, dishwasher and refrigerator. One masonry fireplace.	9.95	2.45	12.40
8. Mechanical	Oil fired forced hot air heat with central air conditioning. Two full baths. Stainless steel kitchen sink with disposal.	4.81	3.08	7.89
9. Electrical	200 amp service, branch circuit wiring with romex cable. Exterior and interior lighting fixtures, receptacles and switches.	2.59	1.44	4.03
10. Overhead	Contractor's overhead and profit.	3.27	2.11	5.38
	Total Cost per Square Foot	$49.99	$32.32	$82.31

To purchase a full set of Sepias, Bill of Materials and Detailed Costs — turn to page 267.

RM-2606A

1 Story
Tudor

1499 Square Feet
3 Bedrooms
2½ Baths
Schedule A

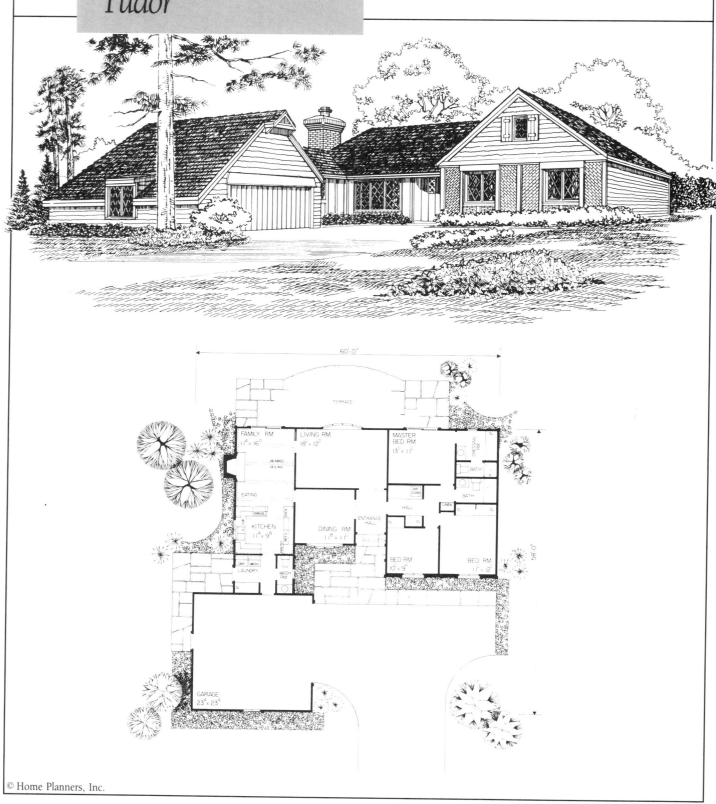

RM-2606A
Cost Estimate

Cost at a Glance

Cost per Square Foot: $93.53
Total Cost: $140,201

Cost by Category

		Cost per Square Foot of Living Area		
		Materials	Installation	Total
1. Site Work	Excavation for the slab and footings.		.45	.45
2. Foundation	Main house and garage — 8″ x 42″ and 12″ x 42″ reinforced concrete trench footings. Slabs — 4″ thick steel trowel finished reinforced concrete over compacted gravel.	3.71	4.68	8.39
3. Framing	Main house — 2 x 6 studs, 16″ on center with 1/2″ plywood sheathing. Garage — 2 x 4 studs, 16″ on center with 1/2″ plywood sheathing. Roof — pre-engineered trusses and site cut 2 x 10 rafters with 1/2″ plywood sheathing.	7.17	4.77	11.94
4. Exterior Walls	Masonry veneer and stucco highlights with wavy edge beveled cedar siding over 15# felt vapor barrier. R-19 and R-11 insulation. Vinyl clad fixed and casement windows, board style entry and sliding glass doors.	17.22	9.33	26.55
5. Roofing	Heavyweight three tab asphalt shingles over 30# felt roofing paper. Aluminum gutters, downspouts, drip edge and flashings.	2.52	1.95	4.47
6. Interiors	Walls and ceilings — 1/2″ and 5/8″ taped and finished gypsum wallboard, primed and painted with one coat latex. Pine interior trim. Flooring — 73% carpet, 11% vinyl, 10% hardwood and 6% ceramic tile.	7.39	6.48	13.87
7. Specialties	Hardwood faced particle board case kitchen cabinets and bathroom vanities with plastic laminate countertops. Washer, dryer, cooktop with hood, double ovens, dishwasher and refrigerator. Masonry fireplace.	7.25	2.52	9.77
8. Mechanical	Oil fired forced hot air heat with central air conditioning. One full and one 3/4 and one 1/2 bath. Stainless steel double bowl kitchen sink with disposal.	4.61	3.09	7.70
9. Electrical	200 amp service, branch circuit wiring with romex cable. Exterior and interior lighting fixtures, receptacles and switches.	2.85	1.42	4.27
10. Overhead	Contractor's overhead and profit.	3.69	2.43	6.12
Total Cost per Square Foot		$56.41	$37.12	$93.53

To purchase a full set of Sepias, Bill of Materials and Detailed Costs — turn to page 267.

RM-2622

2 Story Colonial

1248 Square Feet
3 Bedrooms
2½ Baths
Schedule A

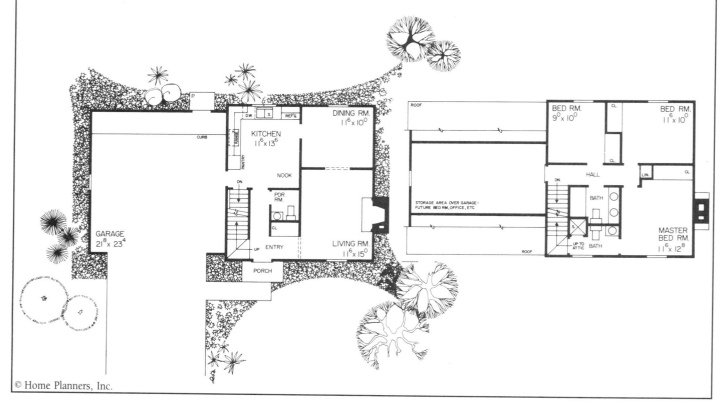

RM-2622
Cost Estimate

Cost at a Glance

Cost per Square Foot: $91.93
Total Cost: $114,728

Cost by Category		Cost per Square Foot of Living Area		
		Materials	Installation	Total
1. Site Work	Excavation for the basement and footings.		.97	.97
2. Foundation	Main house — 12″ wide concrete masonry unit foundation wall on 20″ x 10″ reinforced concrete perimeter footings. Trench footings — 8″ wide reinforced concrete. Slabs — 4″ thick reinforced steel trowel finished concrete over compacted gravel.	3.91	5.30	9.21
3. Framing	Exterior walls — 2 x 6 studs, 16″ on center with 1/2″ plywood sheathing. Garage — 2 x 4 studs, 16″ on center. Floors — 2 x 8 floor joists, 16″ on center with 3/4″ plywood subfloor. Roof — site cut rafters with 1/2″ plywood sheathing.	10.49	7.44	17.93
4. Exterior Walls	Beveled cedar siding over 15# felt vapor barrier with R-19 and R-11 wall insulation. Vinyl clad casement, and double hung windows.	12.56	5.73	18.29
5. Roofing	Three tab asphalt shingles over 30# felt roofing paper. Aluminum gutters, downspouts, drip edge and flashings.	1.40	1.21	2.61
6. Interiors	Walls and ceilings — 1/2″ and 5/8″ taped and finished gypsum wallboard, primed and painted with one coat latex. Pine interior trim, with one coat paint. Flooring — 80% carpet, 13% vinyl and 7% hardwood.	7.96	6.65	14.61
7. Specialties	Hardwood faced particle board case kitchen cabinets and bathroom vanities with plastic laminate countertops. Washer, dryer, range with hood, dishwasher, and refrigerator. One masonry fireplace.	6.30	2.58	8.88
8. Mechanical	Oil fired forced hot air heat. One full bath, one 3/4 bath and one 1/2 bath. Stainless steel double bowl kitchen sink with disposal.	5.39	3.83	9.22
9. Electrical	200 amp service, branch circuit wiring with romex cable. Exterior and interior lighting fixtures, receptacles and switches.	2.66	1.53	4.19
10. Overhead	Contractor's overhead and profit.	3.55	2.47	6.02
Total Cost per Square Foot		$54.22	$37.71	$91.93

To purchase a full set of Sepias, Bill of Materials and Detailed Costs — turn to page 267.

1500 to 2000 Square Feet

Plans	Style	Stories	Total SF	Bedrms	Baths	Page
RM3451	Contemporary	1 Story	1560	2	2	52
RM2818B	Contemporary	1 Story	1566	3	2	54
RM3454	Contemporary	1 Story	1699	3	2	56
RM2671	Contemporary	1 Story	1589	3	2½	58
RM2902	Contemporary	1 Story	1632	3	2½	60
RM2948	Southwestern	1 Story	1830	3	2	62
RM3480	Southwestern	1 Story	1845	3	2	64
RM3478	Southwestern	1 Story	1898	3	2	66
RM3431	Southwestern	1 Story	1899	3	2½	68
RM2878	Traditional	1 Story	1521	3	2	70
RM3340	Traditional	1 Story	1611	3	2	72
RM2672	Traditional	1 Story	1717	3	2	74
RM2802	Traditional	1 Story	1729	3	2	76
RM3345	Traditional	1 Story	1738	3	2	78
RM2947	Traditional	1 Story	1830	3	2	80
RM3314	Traditional	1 Story	1951	3	2	82
RM1920A	Traditional	1 Story	1600	3	2½	84
RM2603	Traditional	1 Story	1949	3	2½	86
RM3376	Traditional	1 Story	1999	3	2½	88
RM3569	Transitional	1 Story	1981	3	2½	90
RM2565C	Tudor	1 Story	1540	3	2½	92
RM2682A	Cape Cod	1½ Story	1720	3	2½	94
RM3571	Cape Cod	1½ Story	1747	3	2½	96
RM2661	Cape Cod	1½ Story	1797	3	2½	98
RM3501	Colonial	1½ Story	1693	3	2½	100
RM3444	Transitional	1½ Story	1973	3	2½	102
RM3379	Colonial	2 Story	1988	3	2½	104
RM2711	Contemporary	2 Story	1999	3	2½	106
RM3316	Traditional	2 Story	1997	3	2½	108
RM1956A	Traditional	2 Story	1718	4	2½	110
RM2488	Tudor	2 Story	1656	3	2	112
RM3331	Tudor	2 Story	1805	3	2	114
RM2974	Victorian	2 Story	1772	3	2½	116
RM3385	Victorian	2 Story	1996	4	2½	118
RM2608	Traditional	Multilevel	1912	4	2½	120

RM-3451

1 Story
Contemporary

1560 Square Feet
2 Bedrooms
2 Baths
Schedule B

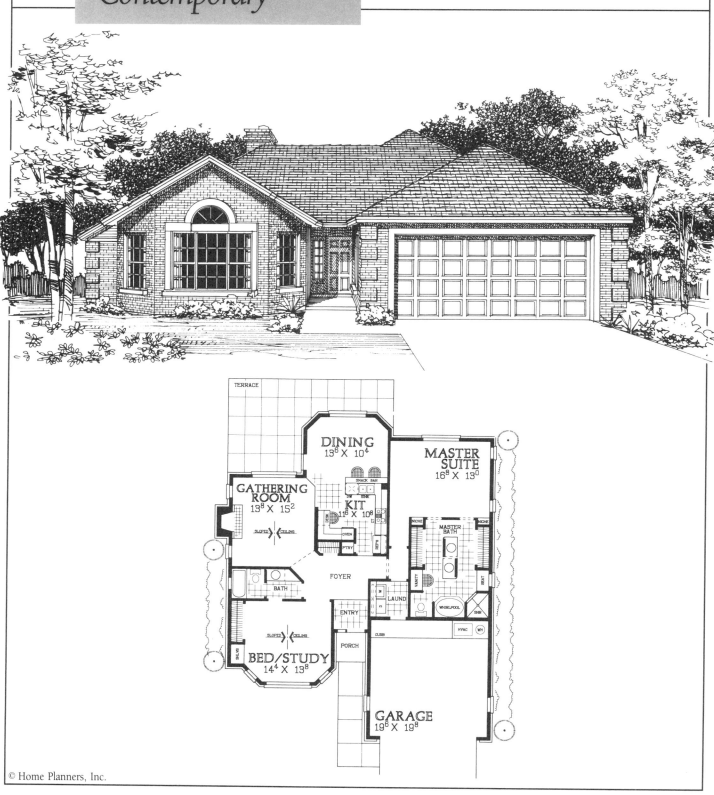

TERRACE

DINING
13⁶ X 10⁴

MASTER SUITE
16⁸ X 13⁰

GATHERING ROOM
13⁸ X 15²

SLOPED CEILING

SNACK BAR

KIT
11⁶ X 10⁸

OVEN

PTRY

NICHE   NICHE

MASTER BATH

VANITY

FOYER

BATH

LAUND

WHIRLPOOL   SHR

SLOPED CEILING

ENTRY

PORCH

HVAC   WH

BED/STUDY
14⁴ X 13⁸

GARAGE
19⁶ X 19⁸

RM-3451
Cost Estimate

Cost at a Glance

Cost per Square Foot: $97.07
Total Cost: $151,429

Cost by Category

		Cost per Square Foot of Living Area		
		Materials	Installation	Total
1. Site Work	Excavation for the slab and footings.		.44	.44
2. Foundation	Main house and garage—10″ wide reinforced concrete foundation wall on 16″ x 10″ reinforced concrete perimeter footings. Slabs—4″ thick steel trowel finished reinforced concrete over compacted gravel.	2.83	3.76	6.59
3. Framing	Exterior walls—2 x 6 studs, 16″ on center with 1/2″ plywood sheathing. Roof—pre-engineered trusses and site cut 2 x 6 rafters with 1/2″ plywood sheathing.	12.09	14.00	26.09
4. Exterior Walls	Brick veneer over 15# felt vapor barrier with R-19 and R-11 insulation. Vinyl clad fixed, double hung and casement windows. Raised panel front entry door, with sidelight.	10.39	7.46	17.85
5. Roofing	Heavyweight three tab asphalt shingles over 30# felt roofing paper. Aluminum gutters, downspouts, drip edge and flashings.	2.28	1.88	4.16
6. Interiors	Walls and ceilings—1/2″ and 5/8″ taped and finished gypsum wallboard, primed and painted with one coat latex. Pine interior trim with one coat paint or stain. Flooring—57% carpet, and 43% vinyl.	6.96	6.27	13.23
7. Specialties	Hardwood faced particle board case kitchen cabinets and bathroom vanities with plastic laminate countertops. Washer, dryer, cooktop with hood, double ovens, dishwasher and refrigerator. Pre-manufactured fireplace.	7.45	1.87	9.32
8. Mechanical	Oil fired forced hot air heat with central air conditioning. One full bath and a master suite with a whirlpool and shower. Stainless steel kitchen sink with disposal.	5.38	3.24	8.62
9. Electrical	200 amp service, branch circuit wiring with romex cable. Exterior and interior lighting fixtures, receptacles and switches.	2.82	1.60	4.42
10. Overhead	Contractor's overhead and profit.	3.51	2.84	6.35
	Total Cost per Square Foot	$53.71	$43.36	$97.07

To purchase a full set of Sepias, Bill of Materials and Detailed Costs—turn to page 267.

RM-2818B

1 Story
Contemporary

1566 Square Feet
3 Bedrooms
2 Baths
Schedule B

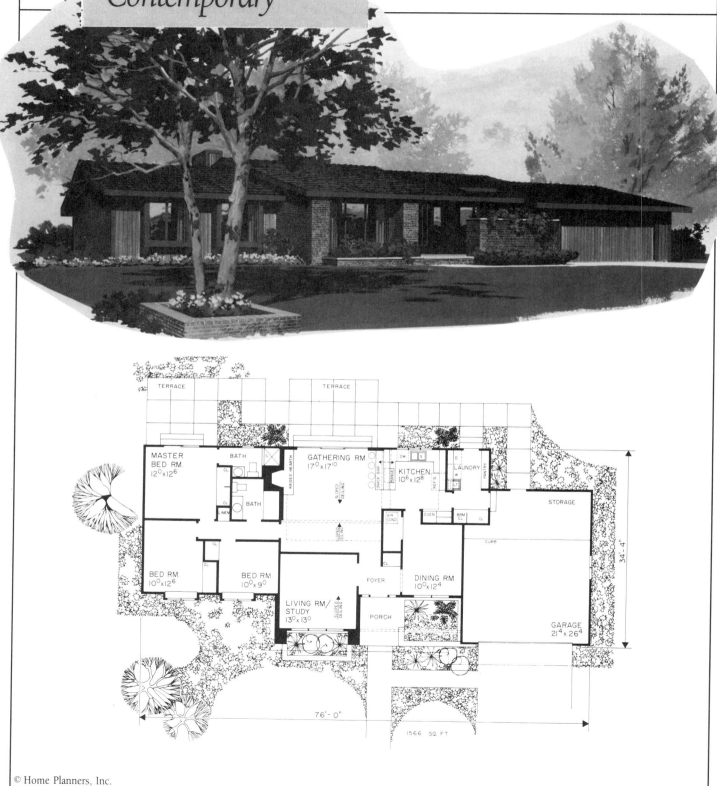

MASTER BED RM. 12⁰x12⁶

BATH

GATHERING RM. 17⁰x17¹⁰

KITCHEN 10⁶x12⁸

LAUNDRY

PANTRY

RAISED HEARTH

SLOPED CEILING

SNACK BAR

RANGE

DW

S

D

REF'G

W

STORAGE

BATH

CL

CL

LINEN

AIR COND

OVEN

BRM CL

CL

CURB

BED RM. 10⁰x12⁶

CL

CL

BED RM. 10⁰x9⁰

CL

LIVING RM./ STUDY 13⁰x13⁰

SLOPED CEILING

FOYER

PORCH

DINING RM. 10⁰x12⁴

GARAGE 21⁴x26⁴

TERRACE

TERRACE

34'-4"

76'-0"

1566 SQ. FT

RM-2818B
Cost Estimate

Cost at a Glance

Cost per Square Foot: $88.92
Total Cost: $139,248

Cost by Category

		Cost per Square Foot of Living Area		
		Materials	Installation	Total
1. Site Work	Excavation for the basement and footings.		1.20	1.20
2. Foundation	Main house — 8″ and 12″ wide concrete masonry unit foundation walls on 16″ x 10″ reinforced concrete perimeter footings. Trench footings — 8″ wide reinforced concrete. Slabs — 4″ thick steel trowel finished reinforced concrete over compacted gravel.	5.25	6.82	12.07
3. Framing	Exterior walls — 2 x 6 studs, 16″ on center with 1/2″ plywood sheathing. Garage — 2 x 4 studs, 16″ on center. Floor — 2 x 10 joists, 16″ on center with 3/4″ plywood subfloor. Roof — site cut rafters with 1/2″ plywood sheathing.	10.59	6.27	16.86
4. Exterior Walls	Brick veneer and Texture 1-11 siding over 15# felt vapor barrier with R-19 and R-11 insulation. Vinyl clad casement windows and sliding glass patio doors.	11.29	4.54	15.83
5. Roofing	Heavyweight three tab asphalt shingles over 30# felt roofing paper. Aluminum gutters, downspouts, drip edge and flashings.	1.73	1.62	3.35
6. Interiors	Walls and ceilings — 1/2″ and 5/8″ taped and finished gypsum wallboard, primed and painted with one coat latex. Pine interior trim with one coat paint or stain. Flooring — 68% carpet, 15% vinyl, 12% hardwood and 5% ceramic tile.	6.90	5.63	12.53
7. Specialties	Hardwood faced particle board case kitchen cabinets and bathroom vanities with plastic laminate countertops. Washer, dryer, cooktop with hood, double wall oven, dishwasher and refrigerator. One pre-fabricated fireplace.	7.47	2.26	9.73
8. Mechanical	Oil fired forced hot air heat with central air conditioning. One full bath and one 3/4 bath. Stainless steel kitchen sink with disposal.	4.31	2.91	7.22
9. Electrical	200 amp service, branch circuit wiring with romex cable. Exterior and interior lighting fixtures, receptacles and switches.	2.73	1.58	4.31
10. Overhead	Contractor's overhead and profit.	3.52	2.30	5.82
	Total Cost per Square Foot	$53.79	$35.13	$88.92

To purchase a full set of Sepias, Bill of Materials and Detailed Costs — turn to page 267.

RM-3454

1 Story Contemporary

1699 Square Feet
3 Bedrooms
2 Baths
Schedule B

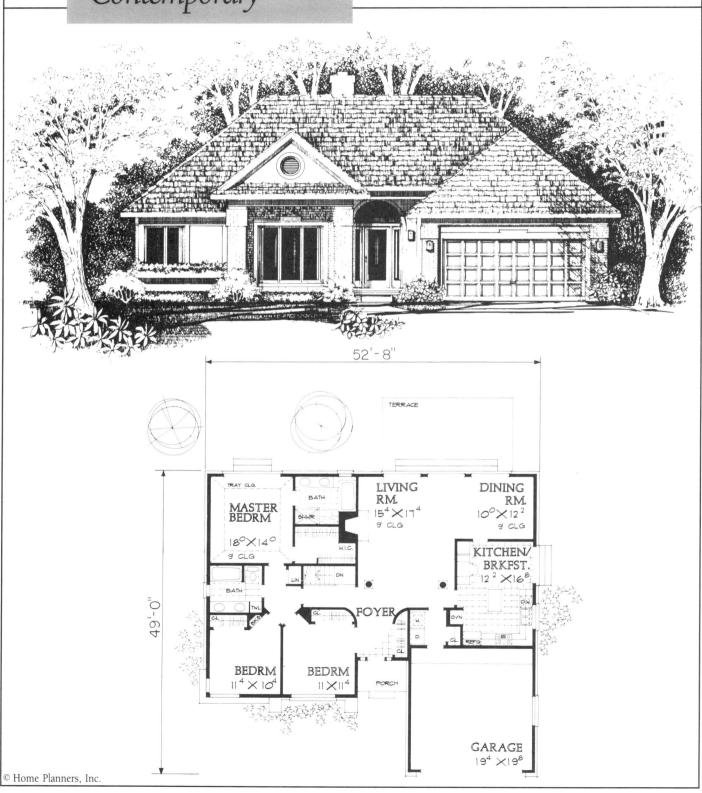

52'-8"

49'-0"

TERRACE

TRAY CLG

MASTER BEDRM
18⁰ X 14⁰
9' CLG

BATH
SHWR
W.I.C.

LIVING RM.
15⁴ X 17⁴
9' CLG

DINING RM.
10⁰ X 12²
9' CLG

KITCHEN/ BRKFST.
12² X 16⁸

BATH
TWL
CL

LIN
DN

FOYER

CL

OVN
REF'G

BEDRM
11⁴ X 10⁴

BEDRM
11 X 11⁴

PORCH

W.
D.
CL

GARAGE
19⁴ X 19⁸

RM-3454
Cost Estimate

Cost at a Glance

Cost per Square Foot: *$89.93*
Total Cost: *$152,791*

Cost by Category

		Cost per Square Foot of Living Area		
		Materials	Installation	Total
1. Site Work	Excavation for the basement and footings.		1.16	1.16
2. Foundation	Main house—10″ wide reinforced concrete foundation wall on 20″ x 10″ reinforced concrete perimeter footings. Trench footings—8″ and 10″ wide reinforced concrete. Slabs—4″ thick steel trowel finished reinforced concrete over compacted gravel.	4.93	6.14	11.07
3. Framing	Exterior walls—2 x 6 studs, 16″ on center with 1/2″ plywood sheathing. Garage—2 x 4 studs, 16″ on center. Floor—2 x 10 floor joists, 16″ on center with 3/4″ plywood subfloor. Roof—pre-engineered trusses and site cut rafters with 1/2″ plywood sheathing.	9.39	6.53	15.92
4. Exterior Walls	Brick veneer over 15# felt vapor barrier with R-19 and R-11 insulation. Vinyl clad casement windows and sliding glass patio doors.	11.52	5.87	17.39
5. Roofing	Heavyweight three tab asphalt shingles over 30# felt roofing paper. Aluminum gutters, downspouts, drip edge and flashings.	1.98	1.85	3.83
6. Interiors	Walls and ceilings—1/2″ and 5/8″ taped and finished gypsum wallboard, primed and painted with one coat latex. Pine interior trim with one coat paint or stain. Flooring—70% carpet, 18% vinyl and 12% ceramic tile.	7.26	6.42	13.68
7. Specialties	Hardwood faced particle board case kitchen cabinets and bathroom vanities with plastic laminate countertops. Washer, dryer, cooktop with hood, double ovens, dishwasher and refrigerator. One pre-manufactured fireplace.	7.51	1.82	9.33
8. Mechanical	Oil fired forced hot air heat with central air conditioning. One full bath and a master suite with a tub and shower. Stainless steel kitchen sink with disposal.	4.55	3.06	7.61
9. Electrical	200 amp service, branch circuit wiring with romex cable. Exterior and interior lighting fixtures, receptacles and switches.	2.64	1.42	4.06
10. Overhead	Contractor's overhead and profit.	3.48	2.40	5.88
	Total Cost per Square Foot	$53.26	$36.67	$89.93

To purchase a full set of Sepias, Bill of Materials and Detailed Costs—turn to page 267.

RM-2671
1 Story Contemporary

1589 Square Feet
3 Bedrooms
2½ Baths
Schedule B

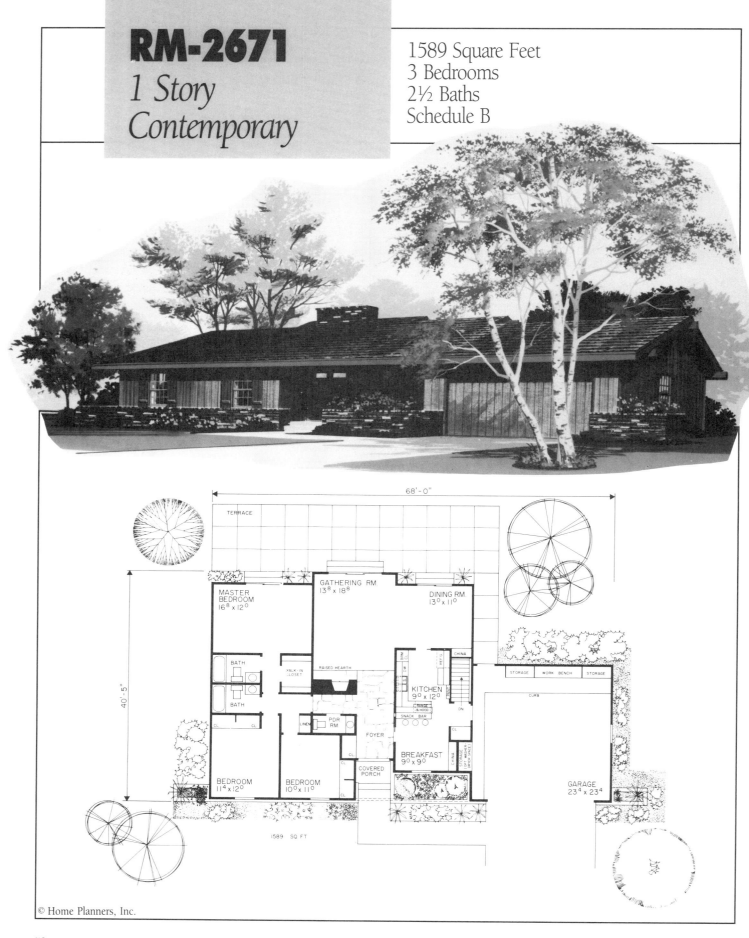

TERRACE

68'-0"

40'-5"

MASTER BEDROOM
16⁸ x 12⁰

GATHERING RM.
13⁸ x 18⁸

DINING RM.
13⁰ x 11⁰

BATH

WALK-IN CLOSET

RAISED HEARTH

CHINA

KITCHEN
9⁰ x 12⁰

BATH

STORAGE   WORK BENCH   STORAGE

CURB

RANGE & HOOD

SNACK BAR

CL

LINEN

PDR RM

FOYER

BREAKFAST
9⁰ x 9⁰

STORAGE

CHINA

CL

BEDROOM
11⁴ x 12⁰

BEDROOM
10⁰ x 11⁰

COVERED PORCH

GARAGE
23⁴ x 23⁴

1589 SQ FT

RM-2671
Cost Estimate

Cost at a Glance

Cost per Square Foot: $91.06
Total Cost: $144,694

Cost by Category

		Cost per Square Foot of Living Area		
		Materials	Installation	Total
1. Site Work	Excavation for the basement and footings.		1.19	1.19
2. Foundation	Main house—12″ wide concrete masonry unit foundation wall on 20″ x 10″ reinforced concrete perimeter footings. Trench footings—8″ and 12″ wide reinforced concrete. Slabs—4″ thick reinforced steel trowel finished concrete over compacted gravel.	5.52	7.24	12.76
3. Framing	Exterior walls—2 x 6 studs, 16″ on center with 1/2″ plywood sheathing. Garage—2 x 4 studs, 16″ on center. Floors—2 x 10 floor joists, 16″ on center with 3/4″ plywood subfloor. Roof—pre-engineered trusses with 1/2″ plywood sheathing.	9.75	5.83	15.58
4. Exterior Walls	Texture 1-11 siding over 15# felt vapor barrier with R-19 and R-11 wall insulation. Vinyl clad double hung windows and sliding patio doors.	11.92	4.89	16.81
5. Roofing	Heavyweight three tab asphalt shingles over 30# felt roofing paper. Aluminum gutters, downspouts, drip edge and flashings.	2.22	1.75	3.97
6. Interiors	Walls and ceilings—1/2″ and 5/8″ taped and finished gypsum wallboard, primed and painted with one coat latex. Pine interior trim, with one coat paint or stain. Flooring—70% carpet, 18% vinyl, and 12% ceramic tile.	7.01	6.01	13.02
7. Specialties	Hardwood faced particle board case kitchen cabinets and bathroom vanities with plastic laminate countertops. Washer, dryer, range with hood, dishwasher, and refrigerator. One masonry fireplace.	6.76	3.27	10.03
8. Mechanical	Oil fired forced hot air heat with central air conditioning. Two full baths and one 1/2 bath. Stainless steel double bowl kitchen sink with disposal.	4.59	3.13	7.72
9. Electrical	200 amp service, branch circuit wiring with romex cable. Exterior and interior lighting fixtures, receptacles and switches.	2.63	1.39	4.02
10. Overhead	Contractor's overhead and profit.	3.53	2.43	5.96
Total Cost per Square Foot		$53.93	$37.13	$91.06

To purchase a full set of Sepias, Bill of Materials and Detailed Costs—turn to page 267.

RM-2902

1 Story Contemporary

1632 Square Feet
3 Bedrooms
2½ Baths
Schedule B

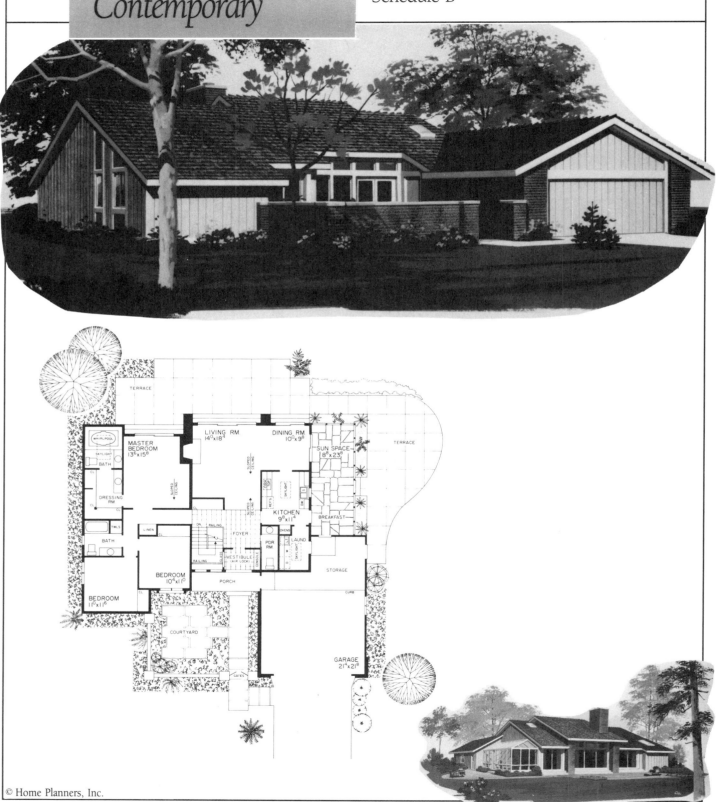

RM-2902
Cost Estimate

Cost per Square Foot: $106.06
Total Cost: $173,089

Cost by Category

		Cost per Square Foot of Living Area		
		Materials	Installation	Total
1. Site Work	Excavation for the basement and footings.		1.18	1.18
2. Foundation	Main house — 12″ concrete block wall on a 20″ x 10″ reinforced concrete footing. Garage and terrace — 8″ and 12″ wide x 42″ reinforced concrete foundation wall. Slabs — 4″ thick reinforced steel trowel finished concrete on 4″ compacted gravel.	6.59	8.28	14.87
3. Framing	Main house — 2 x 6 studs, 16″ on center with 1/2″ plywood sheathing. Garage — 2 x 4 studs, 16″ on center. Floor — 2 x 12 joists, 16″ on center with 3/4″ tongue and groove plywood subfloor. Roof — site cut 2 x 10 and 2 x 8 rafters with 1/2″ plywood sheathing.	11.80	6.92	18.72
4. Exterior Walls	Vertical 1″ x 12″ cedar siding and masonry veneer over 15# felt vapor barrier. Vinyl clad fixed and casement windows and sliding glass doors. Paneled entry doors. R-19 and R-11 insulation.	16.07	6.28	22.35
5. Roofing	Heavyweight three tab asphalt roof shingles on 30# felt paper. Aluminum drip edge, flashings, gutters and downspouts. Dome type skylight.	2.21	1.78	3.99
6. Interiors	Wall finish — one coat primer and one coat paint on 1/2″ or 5/8″ gypsum wallboard. Pine door, window and baseboard moldings. Flooring — 76% carpet, 12% vinyl, 5% ceramic tile and 7% hardwood.	7.32	6.14	13.46
7. Specialties	Hardwood faced, particle board case kitchen cabinets and bath vanities with plastic laminate countertops. Washer, dryer, cooktop with hood, double ovens, dishwasher and refrigerator. A masonry fireplace, courtyard and terrace.	7.22	3.72	10.94
8. Mechanical	Oil fired forced hot air heat with central air conditioning. Two full baths and one half bath. Double bowl kitchen sink with disposal.	5.47	3.41	8.88
9. Electrical	200 amp service, branch circuit wiring with romex cable. Exterior and interior lighting fixtures, receptacles and switches.	3.02	1.71	4.73
10. Overhead	Contractor's overhead and profit.	4.18	2.76	6.94
Total Cost per Square Foot		$63.88	$42.18	$106.06

To purchase a full set of Sepias, Bill of Materials and Detailed Costs — turn to page 267.

RM-2948

*1 Story
Southwestern*

1830 Square Feet
3 Bedrooms
2 Baths
Schedule B

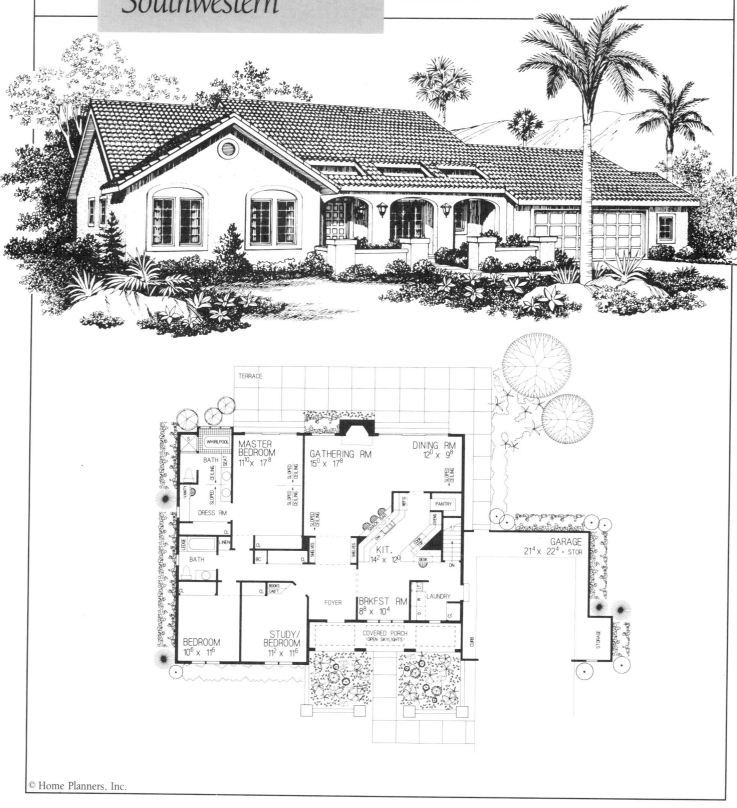

RM-2948
Cost Estimate

Cost per Square Foot: $100.86
Total Cost: $184,573

Cost by Category

		Cost per Square Foot of Living Area		
		Materials	Installation	Total
1. Site Work	Excavation for the basement and footings.		1.12	1.12
2. Foundation	Main house — 10″ wide reinforced concrete foundation wall on 20″ x 10″ reinforced concrete perimeter footings. Trench footings — 8″ wide reinforced concrete. Slabs — 4″ thick reinforced steel trowel finished concrete over compacted gravel.	5.32	6.56	11.88
3. Framing	Exterior walls — 2 x 6 studs, 16″ on center with 1/2″ plywood sheathing. Garage — 2 x 4 studs, 16″ on center. Floors — 2 x 12 floor joists, 16″ on center with 3/4″ plywood subfloor. Roof — pre-engineered trusses and site cut rafters with 3/4″ plywood sheathing.	14.58	9.38	23.96
4. Exterior Walls	1″ thick stucco siding on 3/8″ high rib metal lath over 15# felt vapor barrier with R-19 and R-11 wall insulation. Vinyl clad fixed, and casement windows and sliding glass patio doors.	8.83	2.62	11.45
5. Roofing	Straight barrel mission clay tiles over 30# felt roofing paper. Aluminum gutters, downspouts, drip edge and flashings.	9.02	2.65	11.67
6. Interiors	Walls and ceilings — 1/2″ and 5/8″ taped and finished gypsum wallboard, primed and painted with one coat latex. Pine interior trim. Flooring — 61% carpet, 23% vinyl, 8% hardwood and 8% ceramic tile.	7.21	6.68	13.89
7. Specialties	Hardwood faced particle board case kitchen cabinets and bathroom vanities with plastic laminate countertops. Washer, dryer, cooktop with hood, double ovens, dishwasher, and refrigerator. One masonry fireplace.	6.48	2.28	8.76
8. Mechanical	Oil fired forced hot air heat with central air conditioning. One full bath and a master suite with a whirlpool and shower. Stainless steel double bowl kitchen sink with disposal.	5.01	3.10	8.11
9. Electrical	200 amp service, branch circuit wiring with romex cable. Exterior and interior lighting fixtures, receptacles and switches.	2.27	1.15	3.42
10. Overhead	Contractor's overhead and profit.	4.11	2.49	6.60
Total Cost per Square Foot		$62.83	$38.03	$100.86

To purchase a full set of Sepias, Bill of Materials and Detailed Costs — turn to page 267.

RM-3480

1 Story
Southwestern

1845 Square Feet
3 Bedrooms
2 Baths
Schedule B

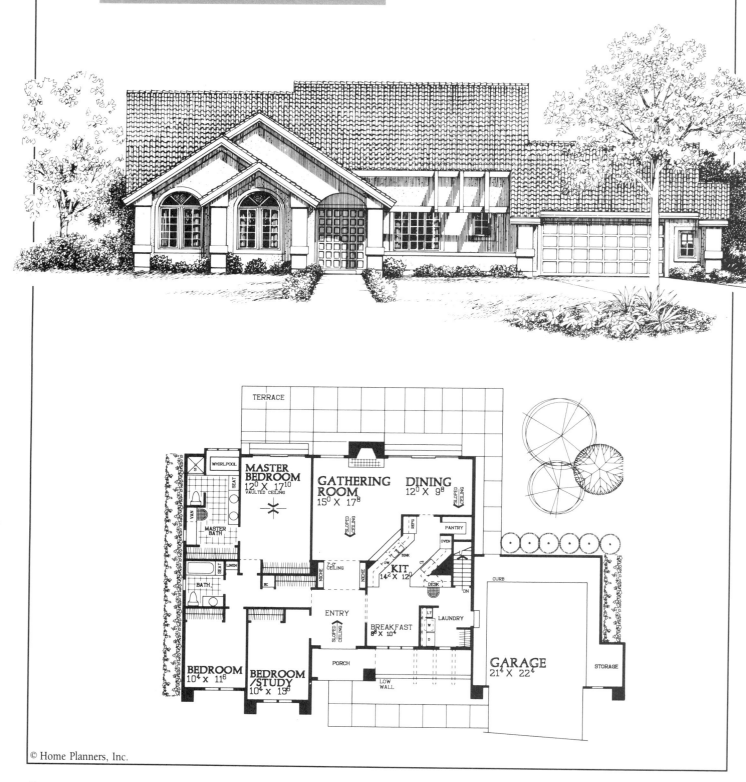

RM-3480
Cost Estimate

Cost at a Glance

Cost per Square Foot: $102.85
Total Cost: $189,758

Cost by Category

		Cost per Square Foot of Living Area		
		Materials	Installation	Total
1. Site Work	Excavation for the basement and footings.		1.12	1.12
2. Foundation	Main house — 10″ wide reinforced concrete foundation wall on 20″ x 10″ reinforced concrete perimeter footings. Trench footings — 8″ wide reinforced concrete. Slabs — 4″ thick reinforced steel trowel finished concrete over compacted gravel.	5.40	6.88	12.28
3. Framing	Exterior walls — 2 x 6 studs, 16″ on center with 1/2″ plywood sheathing. Garage — 2 x 4 studs, 16″ on center. Floors — 2 x 12 floor joists, 16″ on center with 3/4″ plywood subfloor. Roof — pre-engineered trusses and site cut rafters with 3/4″ plywood sheathing.	12.04	8.74	20.78
4. Exterior Walls	1″ stucco siding on 3/8″ high rib metal lath over 15# felt vapor barrier with R-19 and R-11 wall insulation. Vinyl clad fixed, half round and casement windows and sliding glass patio doors.	10.26	3.59	13.85
5. Roofing	Concrete tile roof shingles over 30# felt roofing paper. Aluminum gutters, downspouts, drip edge and flashings.	10.14	3.01	13.15
6. Interiors	Walls and ceilings — 1/2″ and 5/8″ taped and finished gypsum wallboard, primed and painted with one coat latex. Pine interior trim, with one coat paint or stain. Flooring — 55% carpet, 20% vinyl, 12% hardwood and 13% ceramic tile.	6.80	6.56	13.36
7. Specialties	Hardwood faced particle board case kitchen cabinets and bathroom vanities with plastic laminate countertops. Washer, dryer, cooktop with hood, double ovens, dishwasher, and refrigerator. One masonry fireplace.	6.86	2.68	9.54
8. Mechanical	Oil fired forced hot air heat with central air conditioning. One full bath and a master suite with a whirlpool and shower. Stainless steel double bowl kitchen sink with disposal.	5.07	3.11	8.18
9. Electrical	200 amp service, branch circuit wiring with romex cable. Exterior and interior lighting fixtures, receptacles and switches.	2.54	1.32	3.86
10. Overhead	Contractor's overhead and profit.	4.14	2.59	6.73
	Total Cost per Square Foot	$63.25	$39.60	$102.85

To purchase a full set of Sepias, Bill of Materials and Detailed Costs — turn to page 267.

RM-3478

*1 Story
Southwestern*

1898 Square Feet
3 Bedrooms
2 Baths
Schedule B

TERRACE

EATING

HEARTH

FAMILY KITCHEN
25⁴ X 16²

SLOPED ← CEILING

DW

SINK

SNACK BAR

OVEN

MASTER BEDRM
13⁰ X 14¹⁰

SLOPED ↓ CEILING

MASTER BATH

SEAT

LINEN

WALK-IN CLOSET

ARCHED OPENING

NICHE

LINEN

BATH

DINING
11² X 12⁰

CEILING ← SLOPED

FOYER
9'-0" CLG

FLAT ARCH

HALL
9'-0" CLG

ARCHED OPENING

NICHE

LAUNDRY
D **W**

BEDRM
11⁰ X 11⁰
9'-0" CEILING

ARCHED OPENING

SLOPED ← CEILING

STUDY
11⁰ X 10⁸
9'-0" CEILING

WH **HVAC** **STORAGE**

LIVING
12⁰ X 10⁸
• BAY

PORCH

CURB

GARAGE
20⁸ X 24⁰

© Home Planners, Inc.

RM-3478
Cost Estimate

Cost at a Glance

Cost per Square Foot: $94.70
Total Cost: $179,740

Cost by Category

		Cost per Square Foot of Living Area		
		Materials	Installation	Total
1. Site Work	Excavation for the slab and footings.		.37	.37
2. Foundation	Main house — 6″ wide reinforced concrete foundation walls on 16″ x 10″ reinforced concrete perimeter footings. Slabs — 4″ thick steel trowel finished reinforced concrete over compacted gravel.	2.63	3.42	6.05
3. Framing	Exterior walls — 2 x 6 studs, 16″ on center with 1/2″ plywood sheathing. Garage — 2 x 4 studs, 16″ on center. Roof — pre-engineered trusses and site cut rafters with 5/8″ plywood sheathing.	13.00	15.53	28.53
4. Exterior Walls	3 coat stucco system over 15# felt vapor barrier with R-19 and R-11 insulation. Vinyl clad fixed and double hung windows and a sliding glass patio door.	6.56	1.70	8.26
5. Roofing	Mission style clay tile roofing over 30# felt roofing paper. Aluminum gutters, downspouts, drip edge and flashings.	8.30	2.62	10.92
6. Interiors	Walls and ceilings — 1/2″ and 5/8″ taped and finished gypsum wallboard, primed and painted with one coat latex. Pine interior trim with one coat paint or stain. Flooring — 52% carpet, 2% vinyl, and 46% ceramic tile.	7.41	6.74	14.15
7. Specialties	Hardwood faced particle board case kitchen cabinets and bathroom vanities with plastic laminate countertops. Washer, dryer, cooktop with hood, double ovens, dishwasher and refrigerator. One pre-manufactured fireplace.	6.67	1.52	8.19
8. Mechanical	Oil fired forced hot air heat with central air conditioning. One full bath, and a master suite with a whirlpool and shower. Stainless steel kitchen sink with disposal.	4.94	3.03	7.97
9. Electrical	200 amp service, branch circuit wiring with romex cable. Exterior and interior lighting fixtures, receptacles and switches.	2.57	1.49	4.06
10. Overhead	Contractor's overhead and profit.	3.65	2.55	6.20
	Total Cost per Square Foot	$55.73	$38.97	$94.70

To purchase a full set of Sepias, Bill of Materials and Detailed Costs — turn to page 267.

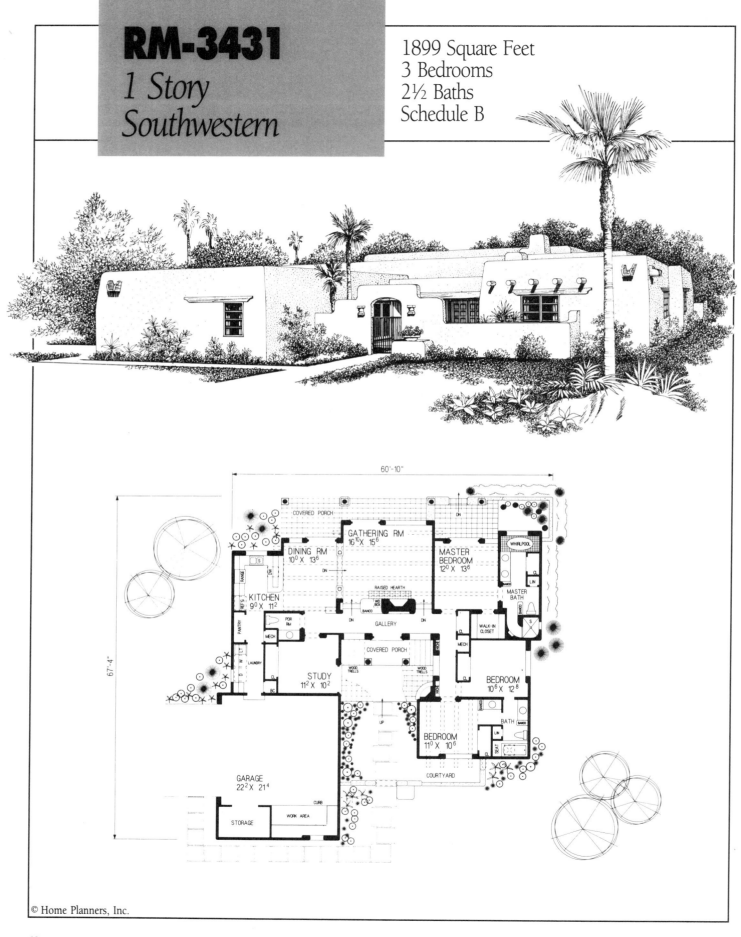

RM-3431

1 Story
Southwestern

1899 Square Feet
3 Bedrooms
2½ Baths
Schedule B

60'-10"

67'-4"

COVERED PORCH

GATHERING RM
16¹⁰ X 15⁶

DINING RM
10⁰ X 13⁶

MASTER
BEDROOM
12⁰ X 13⁶

WHIRLPOOL

MASTER
BATH

LIN

BANCO

RAISED HEARTH

KITCHEN
9⁰ X 11²

PANTRY

RANGE

REF'G

DW

S

WD
BOX

BANCO

GALLERY

WALK-IN
CLOSET

S

POR
RM

MECH

MECH

D W T L T

LAUNDRY

BC

STUDY
11² X 10²

WOOD
TRELLIS

COVERED PORCH

WOOD
TRELLIS

BEDROOM
10⁶ X 12⁸

BANCO

BANCO

BATH

LIN

UP

BEDROOM
11⁰ X 10⁶

SEAT

GARAGE
22² X 21⁴

CURB

COURTYARD

STORAGE

WORK AREA

RM-3431
Cost Estimate

Cost at a Glance

Cost per Square Foot: $101.21
Total Cost: $192,197

Cost by Category

		Cost per Square Foot of Living Area		
		Materials	Installation	Total
1. Site Work	Excavation for the slab and footings.		1.10	1.10
2. Foundation	Main house — 6″ x 24″ and 12″ x 24″ reinforced concrete walls on 10″ x 18″ and 10″ x 24″ reinforced concrete footings. Spread footings and continuous footings at interior bearing points. Slabs — 4″ thick reinforced steel trowel finished concrete on 4″ compacted gravel.	4.76	5.48	10.24
3. Framing	Main house and garage — 2 x 6 studs, 16″ on center with 1/2″ plywood sheathing. Roof — combination of 12″ diameter wood vigas and 2 x 12 joist/rafters with 5/8″ plywood sheathing.	17.17	15.02	32.19
4. Exterior Walls	Three coat stucco on high ribbed metal lath siding. Vinyl clad casement and fixed windows and swinging glazed doors. Paneled front entry door. R-19 and R-11 insulation.	8.97	2.80	11.77
5. Roofing	Built-up roofing with 30# felt building paper. 2′ x 2′ skylight.	.52	.89	1.41
6. Interiors	Wall finish — 1/2″ or 5/8″ gypsum wallboard with one coat primer and one coat finish paint. Pine door, window and baseboard moldings, one coat paint or stain. Flooring — 64% carpet, 15% vinyl, 12% hardwood and 9% ceramic tile.	7.13	6.72	13.85
7. Specialties	Hardwood faced, particle board case kitchen cabinets and bath vanities with plastic laminate countertops. Washer, dryer, range with hood, dishwasher and refrigerator. A fireplace and patio.	8.23	1.72	9.95
8. Mechanical	Oil fired forced hot air heat with central air conditioning. One full bath, one 1/2 bath and a master suite with a whirlpool and shower. Double bowl kitchen sink with disposal.	5.17	3.27	8.44
9. Electrical	200 amp service, branch circuit wiring with romex cable. Exterior and interior lighting fixtures, receptacles and switches.	3.97	1.66	5.63
10. Overhead	Contractor's overhead and profit.	3.92	2.71	6.63
	Total Cost per Square Foot	$59.84	$41.37	$101.21

To purchase a full set of Sepias, Bill of Materials and Detailed Costs — turn to page 267.

RM-2878

1 Story Traditional

1521 Square Feet
3 Bedrooms
2 Baths
Schedule B

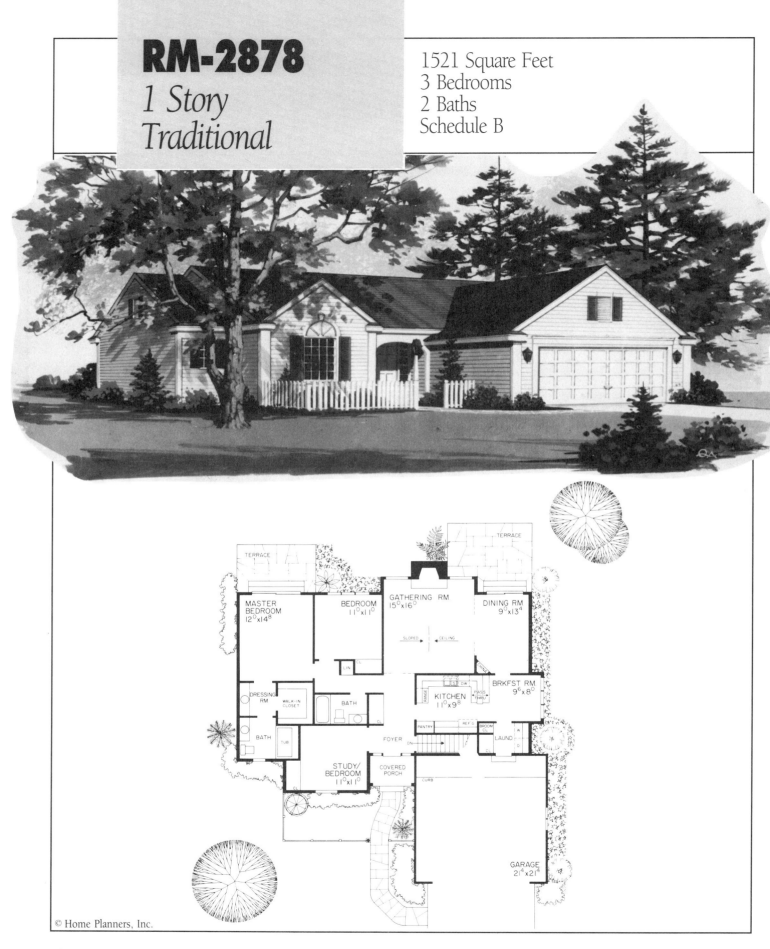

TERRACE

TERRACE

MASTER BEDROOM
12⁰x14⁸

BEDROOM
11⁰x11⁰

GATHERING RM
15⁰x16⁰

DINING RM
9⁰x13⁴

SLOPED ← → CEILING

LIN   CL

DRESSING RM

WALK-IN CLOSET

BATH

KITCHEN
11⁰x9⁸

PASS THRU

BRKFST RM
9⁶x8⁰

RANGE

BATH

TUB

PANTRY   REF G   BROOM CL   LAUND

FOYER   DN

STUDY/ BEDROOM
11⁰x11⁰

COVERED PORCH

CURB

GARAGE
21⁴x21⁴

RM-2878
Cost Estimate

Cost at a Glance
Cost per Square Foot: $89.54
Total Cost: $136,190

Cost by Category

		Cost per Square Foot of Living Area		
		Materials	Installation	Total
1. Site Work	Excavation for the basement and footings.		1.21	1.21
2. Foundation	Main house — 12″ wide concrete masonry unit foundation wall on 20″ x 10″ reinforced concrete perimeter footings. Trench footings — 8″ wide reinforced concrete. Slabs — 4″ thick reinforced steel trowel finished concrete over compacted gravel.	4.96	6.68	11.64
3. Framing	Exterior walls — 2 x 6 studs, 16″ on center with 1/2″ plywood sheathing. Garage — 2 x 4 studs, 16″ on center. Floors — 2 x 10 floor joists, 16″ on center with 3/4″ plywood subfloor. Roof — pre-engineered trusses and site cut rafters with 1/2″ plywood sheathing.	8.53	5.55	14.08
4. Exterior Walls	Beveled cedar siding over 15# felt vapor barrier with R-19 and R-11 wall insulation. Vinyl clad fixed, casement, and double hung windows and patio doors.	13.34	4.77	18.11
5. Roofing	Heavyweight three tab asphalt shingles over 30# felt roofing paper. Aluminum gutters, downspouts, drip edge and flashings.	2.55	1.86	4.41
6. Interiors	Walls and ceilings — 1/2″ and 5/8″ taped and finished gypsum wallboard, primed and painted with one coat latex. Pine interior trim, with one coat paint or stain. Flooring — 65% carpet, 18% vinyl, 9% hardwood and 8% ceramic tile.	7.05	5.87	12.92
7. Specialties	Hardwood faced particle board case kitchen cabinets and bathroom vanities with plastic laminate countertops. Washer, dryer, range with hood, dishwasher, and refrigerator. One pre-fabricated fireplace.	7.81	2.37	10.18
8. Mechanical	Oil fired forced hot air heat with central air conditioning. Two full baths. Stainless steel double bowl kitchen sink with disposal.	4.37	2.92	7.29
9. Electrical	200 amp service, branch circuit wiring with romex cable. Exterior and interior lighting fixtures, receptacles and switches.	2.44	1.41	3.85
10. Overhead	Contractor's overhead and profit.	3.57	2.28	5.85
Total Cost per Square Foot		$54.62	$34.92	$89.54

To purchase a full set of Sepias, Bill of Materials and Detailed Costs — turn to page 267.

RM-3340

1 Story
Traditional

1611 Square Feet
3 Bedrooms
2 Baths
Schedule B

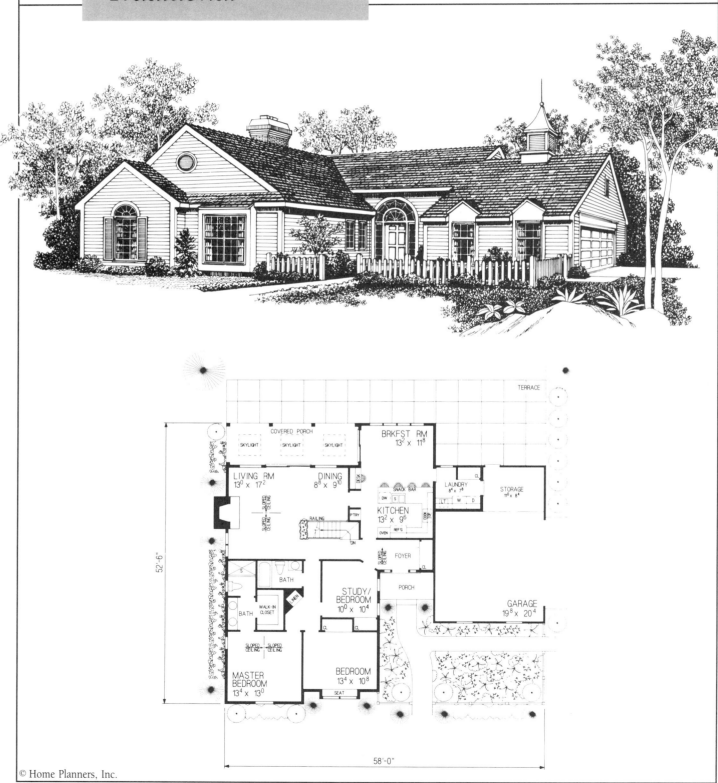

TERRACE

COVERED PORCH

SKYLIGHT SKYLIGHT SKYLIGHT

BRKFST RM
13² x 11⁸

LIVING RM
13⁰ x 17²

DINING
8⁸ x 9¹⁰

DESK

SNACK BAR

LAUNDRY
8⁴ x 7⁸

CL

STORAGE
11⁰ x 8⁴

SLOPED CEILING

SLOPED CEILING

RAILING

P'TRY

DW S

DN

KITCHEN
13² x 9⁶

CL

OVEN REF'G

DOOR W D

52'-6"

BATH

S

BATH

WALK-IN CLOSET

LINEN

STUDY/
BEDROOM
10⁰ x 10⁴

SLOPED CEILING

FOYER

CL

PORCH

GARAGE
19⁸ x 20⁴

SLOPED CEILING SLOPED CEILING

CL CL

BEDROOM
13⁴ x 10⁸

MASTER
BEDROOM
13⁴ x 13⁰

SEAT

58'-0"

RM-3340
Cost Estimate

Cost by Category

Category	Description	Cost per Square Foot of Living Area		
		Materials	Installation	Total
1. Site Work	Excavation for the basement and footings.		1.18	1.18
2. Foundation	Main house — 10" wide reinforced concrete foundation wall on 20" x 10" reinforced concrete perimeter footings. Garage — 8" x 42" reinforced concrete trench footing. Slabs — 4" thick reinforced steel trowel finished concrete over compacted gravel.	5.50	6.89	12.39
3. Framing	Exterior walls — 2 x 6 studs, 16" on center with 1/2" plywood sheathing. Floors — 2 x 10 floor joists, 16" on center with 3/4" plywood subfloor. Roof — pre-engineered trusses and site cut 2 x 8 rafters with 1/2" plywood sheathing.	9.77	6.08	15.85
4. Exterior Walls	Beveled cedar siding over 15# felt vapor barrier with R-19 and R-11 wall insulation. Vinyl clad fixed and casement windows and sliding glass doors. Raised panel front entry door with transom window and sidelights.	15.39	5.76	21.15
5. Roofing	Heavyweight three tab asphalt shingles over 30# felt roofing paper. Aluminum gutters, downspouts, drip edge and flashings.	2.55	2.11	4.66
6. Interiors	Walls and ceilings — 1/2" and 5/8" taped and finished gypsum wallboard, primed and painted with one coat latex. Pine interior trim. Flooring — 65% carpet, 25% vinyl, 7% ceramic tile, and 3% hardwood.	7.57	6.31	13.88
7. Specialties	Hardwood faced particle board case kitchen cabinets and bathroom vanities with plastic laminate countertops. Washer, dryer, cooktop with hood, dishwasher, and refrigerator. Masonry fireplace.	6.49	2.79	9.28
8. Mechanical	Oil fired forced hot air heat with central air conditioning. One full and one 3/4 bath. Stainless steel double bowl kitchen sink with disposal.	4.41	3.02	7.43
9. Electrical	200 amp service, branch circuit wiring with romex cable. Exterior and interior lighting fixtures, receptacles and switches.	2.33	1.34	3.67
10. Overhead	Contractor's overhead and profit.	3.78	2.48	6.26
Total Cost per Square Foot		$57.79	$37.96	$95.75

To purchase a full set of Sepias, Bill of Materials and Detailed Costs — *turn to page 267.*

RM-2672

1 Story
Traditional

1717 Square Feet
3 Bedrooms
2 Baths
Schedule B

74

RM-2672
Cost Estimate

Cost at a Glance

Cost per Square Foot: $90.65
Total Cost: $155,646

Cost by Category

		Cost per Square Foot of Living Area		
		Materials	Installation	Total
1. Site Work	Excavation for the basement and footings.		1.15	1.15
2. Foundation	Main house — 12″ wide masonry foundation wall on reinforced concrete perimeter footings. Trench footings — 8″ wide reinforced concrete. Slabs — 4″ thick reinforced steel trowel finished concrete over compacted gravel.	5.30	6.85	12.15
3. Framing	Exterior walls — 2 x 4 studs, 16″ on center with 1/2″ plywood sheathing. Floors — 2 x 10 floor joists, 16″ on center with 3/4″ plywood subfloor. Roof — pre-engineered trusses and site cut rafters with 1/2″ plywood sheathing.	9.94	5.72	15.66
4. Exterior Walls	Beveled cedar siding over 15# felt vapor barrier with R-13 and R-11 wall insulation. Vinyl clad double hung and casement windows and sliding glass patio doors.	12.25	5.60	17.85
5. Roofing	Heavyweight three tab asphalt shingles over 30# felt roofing paper. Aluminum gutters, downspouts, drip edge and flashings.	1.89	1.59	3.48
6. Interiors	Walls and ceilings — 1/2″ and 5/8″ taped and finished gypsum wallboard, primed and painted with one coat latex. Pine interior trim, with one coat paint or stain. Flooring — 71% carpet, 22% vinyl, and 7% ceramic tile.	7.14	5.79	12.93
7. Specialties	Hardwood faced particle board case kitchen cabinets and bathroom vanities with plastic laminate countertops. Washer, dryer, cooktop with hood, double ovens, dishwasher, and refrigerator. One fireplace.	7.39	3.51	10.90
8. Mechanical	Oil fired forced hot air heat with central air conditioning. One full bath and one 3/4 bath. Stainless steel double bowl kitchen sink with disposal.	4.12	2.78	6.90
9. Electrical	200 amp service, branch circuit wiring with romex cable. Exterior and interior lighting fixtures, receptacles and switches.	2.39	1.31	3.70
10. Overhead	Contractor's overhead and profit.	3.53	2.40	5.93
	Total Cost per Square Foot	$53.95	$36.70	$90.65

To purchase a full set of Sepias, Bill of Materials and Detailed Costs — turn to page 267.

RM-2802

1 Story
Traditional

1729 Square Feet
3 Bedrooms
2 Baths
Schedule B

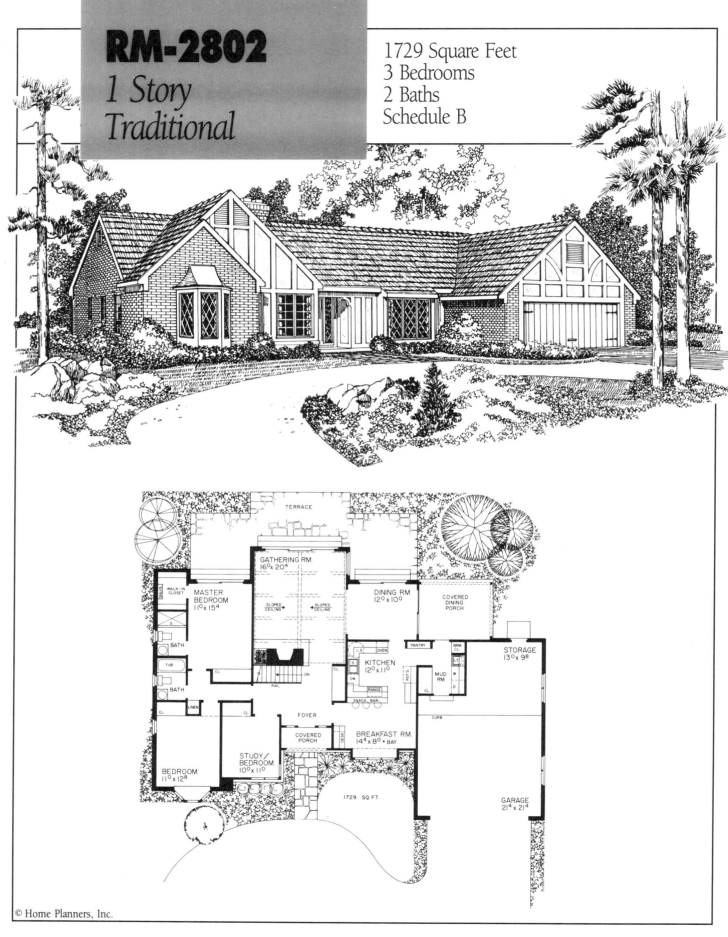

MASTER BEDROOM 11⁰ x 15⁴

WALK-IN CLOSET

BATH

TUB

BATH

LINEN

BEDROOM 11⁰ x 12⁸

STUDY/ BEDROOM 10⁰ x 11⁰

GATHERING RM. 16⁰ x 20⁴

SLOPED CEILING

RAIL

FOYER

COVERED PORCH

DESK

BREAKFAST RM. 14⁴ x 8⁰ + BAY

TERRACE

DINING RM. 12⁰ x 10⁰

COVERED DINING PORCH

KITCHEN 12⁰ x 11⁰

OVEN

RANGE

SNACK BAR

PANTRY

MUD RM.

BRM. CL

STORAGE 13⁰ x 9⁸

CURB

GARAGE 21⁴ x 21⁴

1729 SQ. FT.

RM-2802
Cost Estimate

Cost at a Glance

Cost per Square Foot: $105.26
Total Cost: $181,994

Cost by Category

		Cost per Square Foot of Living Area		
		Materials	Installation	Total
1. Site Work	Excavation for the basement and footings.		1.15	1.15
2. Foundation	Main house — 12″ wide concrete masonry unit foundation wall on 20″ x 10″ reinforced concrete perimeter footings. Trench footings — 8″, 10″ and 12″ wide reinforced concrete. Slabs — 4″ thick steel trowel finished reinforced concrete over compacted gravel.	5.86	7.52	13.38
3. Framing	Exterior walls — 2 x 4 studs, 16″ on center with 1/2″ plywood sheathing. Garage — 2 x 4 studs, 16″ on center. Floor — 2 x 10 joists, 16″ on center with 3/4″ plywood subfloor. Roof — site cut rafters with 5/8″ plywood sheathing.	13.67	8.73	22.40
4. Exterior Walls	Brick veneer, stucco and Texture 1-11 siding over 15# felt vapor barrier with R-19 and R-11 insulation. Vinyl clad fixed and casement windows and sliding glass patio doors.	14.21	7.58	21.79
5. Roofing	Heavyweight three tab asphalt shingles over 30# felt roofing paper. Aluminum gutters, downspouts, drip edge and flashings.	2.07	1.86	3.93
6. Interiors	Walls and ceilings — 1/2″ and 5/8″ taped and finished gypsum wallboard, primed and painted with one coat latex. Pine interior trim. Flooring — 65% carpet, 20% vinyl, 10% hardwood and 5% ceramic tile.	7.48	6.25	13.73
7. Specialties	Hardwood faced particle board case kitchen cabinets and bathroom vanities with plastic laminate countertops. Washer, dryer, cooktop with hood, double wall oven, dishwasher and refrigerator. One masonry fireplace.	7.51	3.87	11.38
8. Mechanical	Oil fired forced hot air heat with central air conditioning. One full bath and one 3/4 bath. Stainless steel kitchen sink with disposal.	4.08	2.71	6.79
9. Electrical	200 amp service, branch circuit wiring with romex cable. Exterior and interior lighting fixtures, receptacles and switches.	2.44	1.39	3.83
10. Overhead	Contractor's overhead and profit.	4.01	2.87	6.88
	Total Cost per Square Foot	$61.33	$43.93	$105.26

*To purchase a full set of Sepias, Bill of Materials and Detailed Costs —**turn to page 267.*

RM-3345

1 Story Traditional

1738 Square Feet
3 Bedrooms
2 Baths
Schedule B

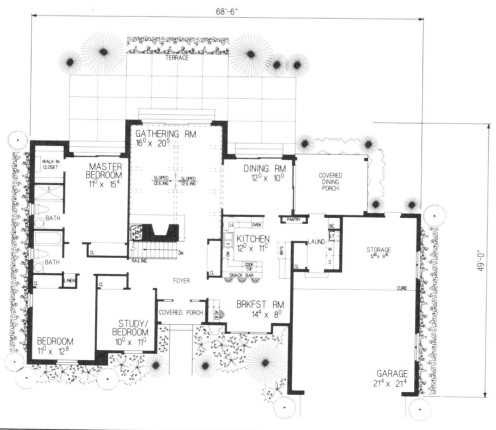

68'-6"

49'-0"

TERRACE

GATHERING RM
16⁰ x 20⁵

WALK-IN CLOSET

MASTER BEDROOM
11⁰ x 15⁴

SLOPED CEILING   SLOPED CEILING

DINING RM
12⁰ x 10⁰

COVERED DINING PORCH

S

BATH

BATH

CL

DN

RAILING

OVEN
LS

KITCHEN
12⁰ x 11⁰

DW

COOK TOP

PANTRY

LAUND

W

D

BC

STORAGE
12⁸ x 9⁰

LINEN

CL

CL

CL

FOYER

SNACK BAR

CURB

BEDROOM
11⁰ x 12⁸

STUDY/ BEDROOM
10⁰ x 11⁰

COVERED PORCH

BRKFST RM
14⁴ x 8⁰

GARAGE
21⁴ x 21⁴

RM-3345
Cost Estimate

Cost at a Glance

Cost per Square Foot: $96.51
Total Cost: $167,734

Cost by Category

		Cost per Square Foot of Living Area		
		Materials	Installation	Total
1. Site Work	Excavation for the basement and footings.		.40	.40
2. Foundation	Main house — 10″ wide reinforced concrete foundation wall on 20″ x 10″ reinforced concrete perimeter footings. Trench footings — 10″ and 12″ wide reinforced concrete. Slabs — 4″ thick steel trowel finished reinforced concrete over compacted gravel.	5.94	7.51	13.45
3. Framing	Exterior walls — 2 x 6 studs, 16″ on center with 1/2″ plywood sheathing. Garage — 2 x 4 studs, 16″ on center. Floor — 2 x 10 joists, 16″ on center with 3/4″ plywood subfloor. Roof — site cut rafters with 1/2″ plywood sheathing.	11.84	7.04	18.88
4. Exterior Walls	Brick veneer, cedar shakes and Texture 1-11 siding over 15# felt vapor barrier with R-19 and R-11 insulation. Vinyl clad fixed, casement and double hung windows and sliding glass patio doors.	13.06	7.14	20.20
5. Roofing	Heavyweight three tab asphalt shingles over 30# felt roofing paper. Aluminum gutters, downspouts, drip edge and flashings.	1.84	1.63	3.47
6. Interiors	Walls and ceilings — 1/2″ and 5/8″ taped and finished gypsum wallboard, primed and painted with one coat latex. Pine interior trim. Flooring — 70% carpet, 20% vinyl, 7% hardwood and 3% ceramic tile.	7.41	6.40	13.81
7. Specialties	Hardwood faced particle board case kitchen cabinets and bathroom vanities with plastic laminate countertops. Washer, dryer, cooktop with hood, double wall ovens, dishwasher and refrigerator. One masonry fireplace.	6.29	2.92	9.21
8. Mechanical	Oil fired forced hot air heat with central air conditioning. One full bath and one 3/4 bath. Stainless steel kitchen sink with disposal.	4.05	2.73	6.78
9. Electrical	200 amp service, branch circuit wiring with romex cable. Exterior and interior lighting fixtures, receptacles and switches.	2.57	1.43	4.00
10. Overhead	Contractor's overhead and profit.	3.71	2.60	6.31
Total Cost per Square Foot		$56.71	$39.80	$96.51

To purchase a full set of Sepias, Bill of Materials and Detailed Costs — turn to page 267.

RM-2947

1 Story Traditional

1830 Square Feet
3 Bedrooms
2 Baths
Schedule B

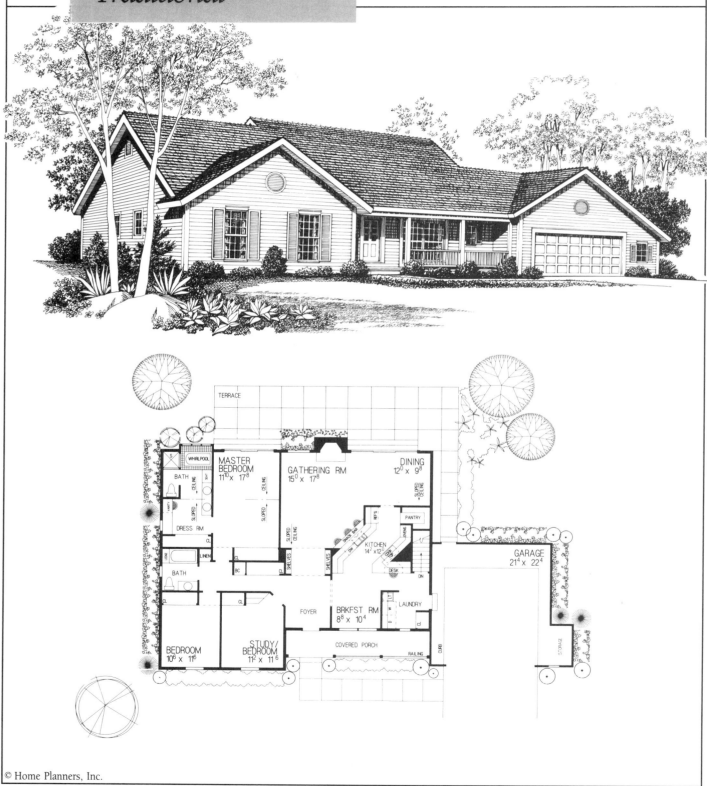

RM-2947
Cost Estimate

Cost at a Glance

Cost per Square Foot: $86.14
Total Cost: $157,636

Cost by Category

		Cost per Square Foot of Living Area		
		Materials	Installation	Total
1. Site Work	Excavation for the basement and footings.		1.12	1.12
2. Foundation	Main house—10″ thick reinforced concrete walls on 18″ x 10″ reinforced concrete footings. Garage—8″ x 54″ reinforced concrete trench walls. Slabs—4″ thick reinforced steel trowel finished concrete on 4″ compacted gravel.	5.02	6.30	11.32
3. Framing	Main house—2 x 6 studs, 16″ on center with 1/2″ plywood sheathing. Garage—2 x 4 studs, 16″ on center. Floor—2 x 10 joists, 16″ on center with 3/4″ tongue and groove plywood subfloor. Roof— pre-engineered trusses and site cut 2 x 6 rafters with 1/2″ plywood sheathing.	9.19	5.65	14.84
4. Exterior Walls	Beveled cedar siding over 15# felt vapor barrier. Vinyl clad fixed, and double hung windows and sliding glass doors. Crossbuck and paneled entry doors. R-19 and R-11 insulation.	11.31	4.09	15.40
5. Roofing	Heavyweight three tab asphalt roof shingles on 30# felt paper. Aluminum drip edge, flashings, gutters and downspouts.	2.43	1.66	4.09
6. Interiors	Wall finish—one coat primer and one coat paint on 1/2″ or 5/8″ gypsum wallboard. Pine door, window and baseboard moldings. Flooring—76% carpet, 15% vinyl, 5% ceramic tile and 4% hardwood.	6.75	6.26	13.01
7. Specialties	Hardwood faced, particle board case kitchen cabinets and bath vanities with plastic laminate countertops. Washer, dryer, cooktop with hood, double ovens, dishwasher and refrigerator. One masonry fireplace, a covered front porch and a terrace.	6.78	2.37	9.15
8. Mechanical	Oil fired forced hot air heat with central air conditioning. One full bath, and a master suite with a tub and shower. Double bowl kitchen sink with disposal.	5.07	3.11	8.18
9. Electrical	200 amp service, branch circuit wiring with romex cable. Exterior and interior lighting fixtures, receptacles and switches.	2.24	1.16	3.40
10. Overhead	Contractor's overhead and profit.	3.41	2.22	5.63
	Total Cost per Square Foot	**$52.20**	**$33.94**	**$86.14**

To purchase a full set of Sepias, Bill of Materials and Detailed Costs—turn to page 267.

RM-3314

1 Story
Traditional

1951 Square Feet
3 Bedrooms
2 Baths
Schedule B

TERRACE

SCREENED PORCH
11⁰ x 10¹⁰

BREAKFAST RM
13⁸ x 11⁴

VERANDA

RAILING

DINING RM
12⁰ x 13⁶

SNACK BAR

REF'G

S   D.W.

KITCHEN
14⁰ x 8⁴

BC

SLOPED CEILING   SLOPED CEILING

MASTER BEDROOM
12⁰ x 15⁰

SLOPED CEILING   SLOPED CEILING

S

WHIRLPOOL

DESK

PANTRY

COOK TOP   OVENS

DN

VANITY

BATH

RAILING

LINEN

WALK-IN CLOSET

WALK-IN CLOSET

SLOPED CEILING   SLOPED CEILING

CL   CL

CL

GATHERING RM
17⁰ x 16⁴

FOYER

BEDROOM
11⁰ x 12⁰ BAY

LINEN

BATH

SEAT

BEDROOM
11⁰ x 12⁰

VERANDA

RAILING

UP

RM-3314
Cost Estimate

Cost at a Glance

Cost per Square Foot: $85.93
Total Cost: $167,649

Cost by Category

Category	Description	Materials	Installation	Total
1. Site Work	Excavation for the basement and footings.		.87	.87
2. Foundation	Main house — 10″ wide reinforced concrete foundation wall on 20″ x 10″ reinforced concrete perimeter footings. Garage — 10″ x 66″ reinforced concrete trench footing. Slabs — 4″ thick reinforced steel trowel finished concrete over compacted gravel.	5.10	6.23	11.33
3. Framing	Exterior walls — 2 x 6 studs, 16″ on center with 1/2″ plywood sheathing. Floors — 2 x 12 floor joists, 16″ on center with 3/4″ plywood subfloor. Roof — pre-engineered trusses with 1/2″ plywood sheathing.	9.50	5.78	15.28
4. Exterior Walls	Cedar shingle siding over 15# felt vapor barrier with R-19 wall insulation. Vinyl clad fixed and casement windows and sliding glass doors. Fully glazed front entry door.	13.13	5.20	18.33
5. Roofing	Heavyweight three tab asphalt shingles over 30# felt roofing paper. Aluminum gutters, downspouts, drip edge and flashings.	1.58	1.30	2.88
6. Interiors	Walls and ceilings — 1/2″ and 5/8″ taped and finished gypsum wallboard, primed and painted with one coat latex. Pine interior trim. Flooring — 68% carpet, 18% vinyl, 6% ceramic tile, and 8% hardwood.	6.27	5.60	11.87
7. Specialties	Hardwood faced particle board case kitchen cabinets and bathroom vanities with plastic laminate countertops. Washer, dryer, cooktop, double ovens, dishwasher, and refrigerator. A masonry fireplace and a veranda.	6.08	2.40	8.48
8. Mechanical	Oil fired forced hot air heat with central air conditioning. One full bath and a master suite with a whirlpool and shower. Stainless steel double bowl kitchen sink with disposal.	4.76	2.87	7.63
9. Electrical	200 amp service, branch circuit wiring with romex cable. Exterior and interior lighting fixtures, receptacles and switches.	2.33	1.31	3.64
10. Overhead	Contractor's overhead and profit.	3.41	2.21	5.62
Total Cost per Square Foot		$52.16	$33.77	$85.93

Column headers: Cost per Square Foot of Living Area — Materials | Installation | Total

To purchase a full set of Sepias, Bill of Materials and Detailed Costs — turn to page 267.

RM-1920A

1 Story
Traditional

1600 Square Feet
3 Bedrooms
2½ Baths
Schedule B

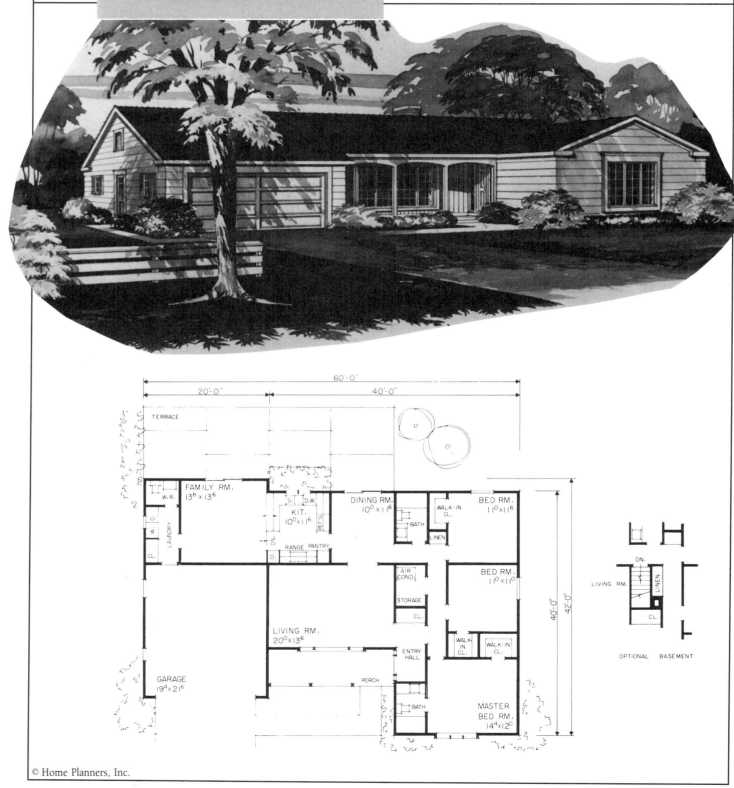

© Home Planners, Inc.

RM-1920A
Cost Estimate

Cost by Category

		Cost per Square Foot of Living Area		
		Materials	Installation	Total
1. Site Work	Excavation for the slab and footings.		.43	.43
2. Foundation	Main house — trench footings — 8″ wide reinforced concrete. Slabs — 4″ thick reinforced steel trowel finished concrete over compacted gravel.	3.23	4.05	7.28
3. Framing	Exterior walls — 2 x 4 studs, 16″ on center with 1/2″ plywood sheathing. Roof — pre-engineered trusses and site cut rafters with 1/2″ plywood sheathing.	5.52	3.67	9.19
4. Exterior Walls	Beveled cedar siding and vertical boards over 15# felt vapor barrier with R-13 and R-11 wall insulation. Vinyl clad fixed, double hung and casement windows and sliding glass patio doors.	8.69	2.97	11.66
5. Roofing	Heavyweight three tab asphalt shingles over 15# felt roofing paper. Aluminum gutters, downspouts, drip edge and flashings.	1.58	1.46	3.04
6. Interiors	Walls and ceilings — 1/2″ and 5/8″ taped and finished gypsum wallboard, primed and painted with one coat latex. Pine interior trim, with one coat paint or stain. Flooring — 66% carpet, 29% vinyl, and 5% ceramic tile. Slate flooring at entry.	6.85	5.85	12.70
7. Specialties	Hardwood faced particle board case kitchen cabinets and bathroom vanities with plastic laminate countertops. Washer, dryer, cooktop with hood, double ovens, dishwasher, and refrigerator.	5.37	1.31	6.68
8. Mechanical	Oil fired forced hot air heat with central air conditioning. One full, one 3/4 and one 1/2 bath. Stainless steel double bowl kitchen sink with disposal.	4.62	3.03	7.65
9. Electrical	200 amp service, branch circuit wiring with romex cable. Exterior and interior lighting fixtures, receptacles and switches.	2.47	1.27	3.74
10. Overhead	Contractor's overhead and profit.	2.68	1.68	4.36
	Total Cost per Square Foot	**$41.01**	**$25.72**	**$66.73**

To purchase a full set of Sepias, Bill of Materials and Detailed Costs — turn to page 267.

RM-2603

1 Story
Traditional

1949 Square Feet
3 Bedrooms
2½ Baths
Schedule B

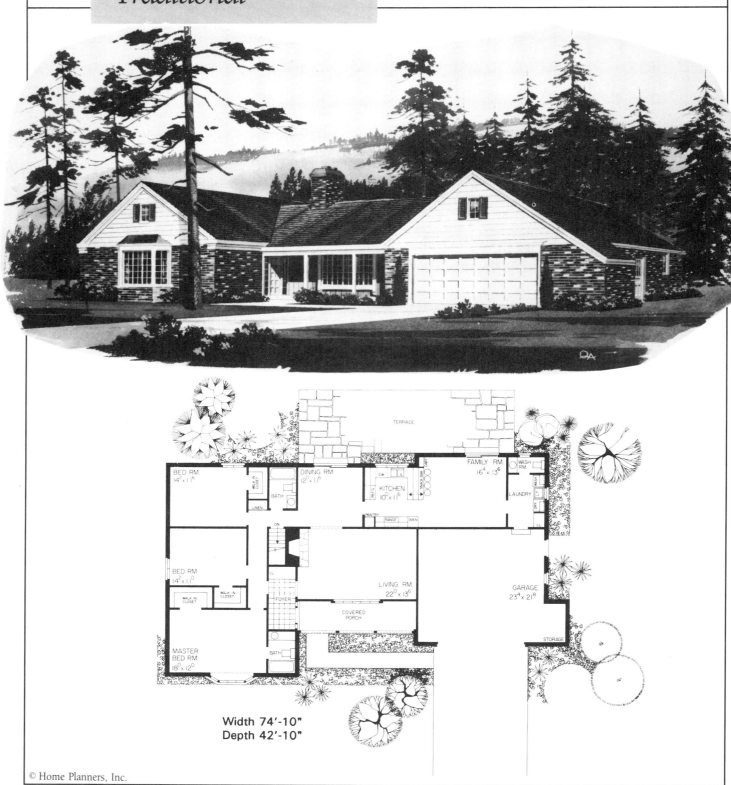

Width 74'-10"
Depth 42'-10"

RM-2603
Cost Estimate

Cost at a Glance

Cost per Square Foot: $83.67
Total Cost: $163,072

Cost by Category

		Cost per Square Foot of Living Area		
		Materials	Installation	Total
1. Site Work	Excavation for the basement and footings.		1.09	1.09
2. Foundation	Main house — 12″ concrete block foundation wall on 20″ x 10″ reinforced concrete perimeter footings. Garage — 8″ x 48″ reinforced concrete trench footings. Slabs — 4″ thick steel trowel finished reinforced concrete over compacted gravel.	5.20	6.74	11.94
3. Framing	Main house — 2 x 6 studs, 16″ on center with 1/2″ plywood sheathing. Floors — 2 x 8 floor joists, 16″ on center with 3/4″ plywood subfloor. Garage — 2 x 4 studs, 16″ on center with 1/2″ plywood sheathing.	8.87	5.28	14.15
4. Exterior Walls	Masonry veneer and beveled cedar siding over 15# felt vapor barrier with R-19 and R-11 insulation. Vinyl clad fixed and casement windows, raised panel entry and sliding glass doors.	11.04	6.32	17.36
5. Roofing	Heavyweight three tab asphalt shingles over 30# felt roofing paper. Aluminum gutters, downspouts, drip edge and flashings. Metal roofing over master bedroom bay window.	1.55	1.44	2.99
6. Interiors	Walls and ceilings — 1/2″ and 5/8″ taped and finished gypsum wallboard, primed and painted with one coat latex. Pine interior trim, painted or stained one coat. Flooring — 65% carpet, 24% vinyl, and 11% ceramic tile.	6.47	5.69	12.16
7. Specialties	Hardwood faced particle board case kitchen cabinets and bathroom vanities with plastic laminate countertops. Washer, dryer, cooktop with hood, double ovens, dishwasher and refrigerator. A masonry fireplace and a front porch.	5.67	2.28	7.95
8. Mechanical	Oil fired forced hot air heat with central air conditioning. Two full and one 1/2 bath. Stainless steel double bowl kitchen sink with disposal.	4.17	2.80	6.97
9. Electrical	200 amp service, branch circuit wiring with romex cable. Exterior and interior lighting fixtures, receptacles and switches.	2.27	1.31	3.58
10. Overhead	Contractor's overhead and profit.	3.17	2.31	5.48
	Total Cost per Square Foot	$48.41	$35.26	$83.67

To purchase a full set of Sepias, Bill of Materials and Detailed Costs — turn to page 267.

RM-3376

1 Story
Traditional

1999 Square Feet
3 Bedrooms
2½ Baths
Schedule B

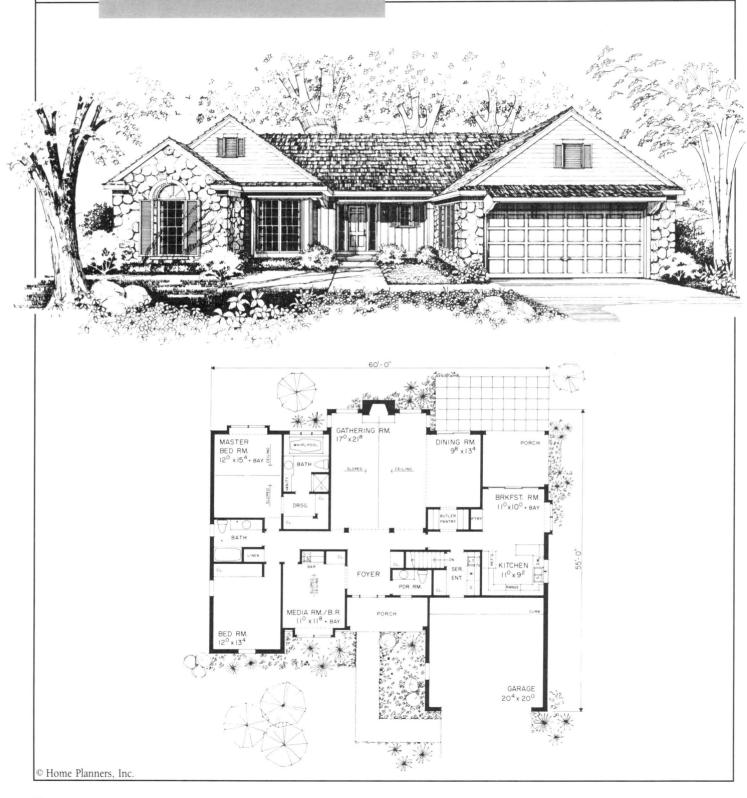

MASTER BED RM.
12⁰ x 15⁴ + BAY

WHIRLPOOL

BATH

VANITY

DRSG.

CL

CL

GATHERING RM.
17⁰ x 21⁸

SLOPED

CEILING

DINING RM.
9⁸ x 13⁴

PORCH

BRKFST. RM.
11⁰ x 10⁰ + BAY

BUTLER PANTRY

P'TRY

BATH

LINEN

CL

CL

BAR

SLOPED CEILING

FOYER

CL

PDR. RM.

SER. ENT.

DN

REF G.

KITCHEN
11⁰ x 9²

RANGE

DW

MEDIA RM./B.R.
11⁰ x 11⁸ + BAY

PORCH

CURB

BED RM.
12⁰ x 13⁴

GARAGE
20⁴ x 20⁰

60'-0"

55'-0"

RM-3376
Cost Estimate

Cost by Category

		Cost per Square Foot of Living Area		
		Materials	Installation	Total
1. Site Work	Excavation for the basement and footings.		1.08	1.08
2. Foundation	Main house—10″ wide reinforced concrete foundation wall on 20″ x 10″ reinforced concrete perimeter footings. Trench footings—8″ and 10″ wide reinforced concrete walls. Slabs—4″ thick steel trowel finished reinforced concrete over compacted gravel.	4.88	6.17	11.05
3. Framing	Exterior walls—2 x 6 studs, 16″ on center with 1/2″ plywood sheathing. Garage—2 x 4 studs, 16″ on center. Floor—2 x 10 joists, 16″ on center with 3/4″ plywood subfloor. Roof—pre-engineered trusses and site cut rafters with 1/2″ plywood sheathing.	9.19	5.58	14.77
4. Exterior Walls	Beveled cedar, stone and brick veneer siding over 15# felt vapor barrier with R-19 and R-11 insulation. Vinyl clad fixed, double hung and casement windows, and a sliding glass patio door.	12.38	5.63	18.01
5. Roofing	Heavyweight three tab asphalt shingles over 30# felt roofing paper. Aluminum gutters, downspouts, drip edge and flashings.	1.72	1.50	3.22
6. Interiors	Walls and ceilings—1/2″ and 5/8″ taped and finished gypsum wallboard, primed and painted with one coat latex. Pine interior trim with one coat paint or stain. Flooring—63% carpet, 21% vinyl, 10% hardwood and 6% ceramic tile.	6.66	5.75	12.41
7. Specialties	Hardwood faced particle board case kitchen cabinets and bathroom vanities with plastic laminate countertops. Washer, dryer, range with hood, dishwasher and refrigerator. One masonry fireplace.	9.80	2.25	12.05
8. Mechanical	Oil fired forced hot air heat with central air conditioning. One full bath, one 1/2 bath and a master suite with a whirlpool and shower. Stainless steel kitchen sink with disposal.	5.21	3.22	8.43
9. Electrical	200 amp service, branch circuit wiring with romex cable. Exterior and interior lighting fixtures, receptacles and switches.	2.02	1.15	3.17
10. Overhead	Contractor's overhead and profit.	3.63	2.26	5.89
Total Cost per Square Foot		$55.49	$34.59	$90.08

To purchase a full set of Sepias, Bill of Materials and Detailed Costs—turn to page 267.

RM-3569

1 Story Transitional

1981 Square Feet
3 Bedrooms
2½ Baths
Schedule B

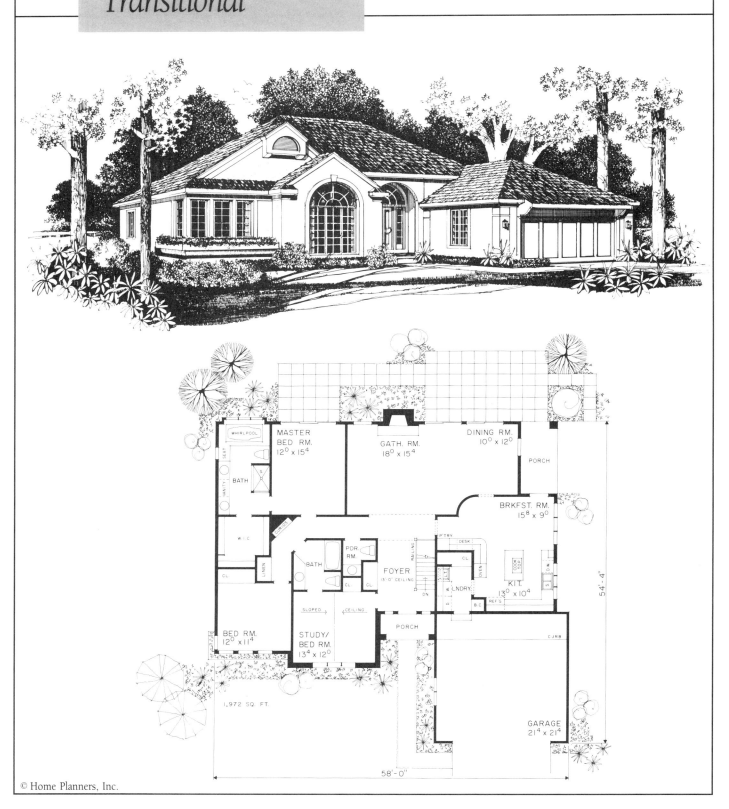

MASTER BED RM. 12⁰ x 15⁴

WHIRLPOOL

SEAT

VANITY

BATH

S.

W.I.C.

LINEN

CL.

PDR. RM.

BATH

CL.

CL.

GATH. RM. 18⁰ x 15⁴

DINING RM. 10⁰ x 12⁰

PORCH

BRKFST. RM. 15⁸ x 9⁰

P'TRY

DESK

CL.

OVEN

RAILING

FOYER 13'0" CEILING

DN

LNDRY.

KIT. 13⁰ x 10⁴

COOK TOP

D.W.

REF'G.

B.C.

BED RM. 12⁰ x 11⁴

STUDY/ BED RM. 13⁴ x 12⁰

SLOPED →

CEILING

PORCH

GARAGE 21⁴ x 21⁴

CJRB

1,972 SQ. FT.

54'-4"

58'-0"

RM-3569
Cost Estimate

Cost at a Glance

Cost per Square Foot: $86.51
Total Cost: $171,376

Cost by Category

		Cost per Square Foot of Living Area		
		Materials	Installation	Total
1. Site Work	Excavation for the basement and footings.		1.08	1.08
2. Foundation	Main house—10″ wide reinforced concrete foundation wall on 16″ x 8″ reinforced concrete perimeter footings. Trench footings—10″ wide reinforced concrete. Slabs—4″ thick reinforced steel trowel finished concrete over compacted gravel.	4.25	5.43	9.68
3. Framing	Exterior walls—2 x 6 studs, 16″ on center with 1/2″ plywood sheathing. Garage—2 x 4 studs, 16″ on center. Floors—2 x 12 floor joists, 16″ on center with 3/4″ plywood subfloor. Roof—pre-engineered trusses and site cut rafters with 1/2″ plywood sheathing.	9.09	5.97	15.06
4. Exterior Walls	Exterior insulation finish system with R-19 and R-11 wall insulation. Vinyl clad fixed, and casement windows and sliding glass patio doors.	11.30	2.70	14.00
5. Roofing	Heavyweight three tab asphalt shingles over 30# felt roofing paper. Aluminum gutters, downspouts, drip edge and flashings.	1.45	1.55	3.00
6. Interiors	Walls and ceilings—1/2″ and 5/8″ taped and finished gypsum wallboard, primed and painted with one coat latex. Pine interior trim, with one coat paint or stain. Flooring—62% carpet, 22% vinyl, 5% hardwood, and 11% ceramic tile.	6.78	5.95	12.73
7. Specialties	Hardwood faced particle board case kitchen cabinets and bathroom vanities with plastic laminate countertops. Washer, dryer, cooktop with hood, double ovens, dishwasher, and refrigerator. One pre-fabricated fireplace.	11.01	1.88	12.89
8. Mechanical	Oil fired forced hot air heat with central air conditioning. One full bath, one 1/2 bath and a master suite with a whirlpool and shower. Stainless steel double bowl kitchen sink with disposal.	5.14	3.25	8.39
9. Electrical	200 amp service, branch circuit wiring with romex cable. Exterior and interior lighting fixtures, receptacles and switches.	2.64	1.38	4.02
10. Overhead	Contractor's overhead and profit.	3.62	2.04	5.66
	Total Cost per Square Foot	$55.28	$31.23	$86.51

To purchase a full set of Sepias, Bill of Materials and Detailed Costs—turn to page 267.

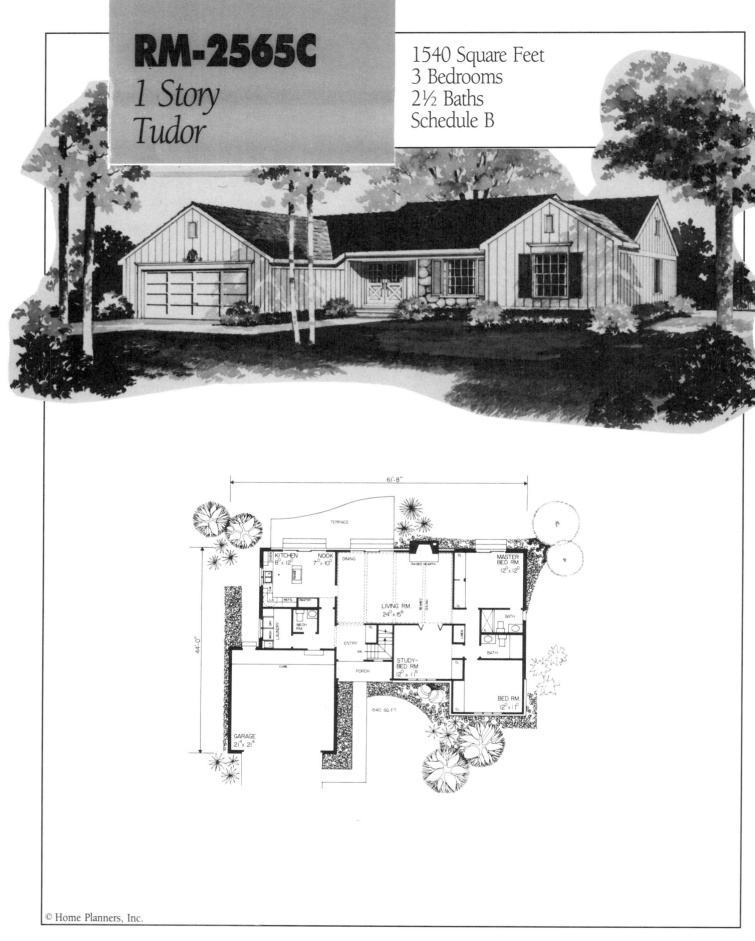

RM-2565C

1 Story
Tudor

1540 Square Feet
3 Bedrooms
2½ Baths
Schedule B

61'-8"

44'-0"

TERRACE

KITCHEN
8⁰ x 12²

NOOK
7⁰ x 10⁰

DINING

RAISED HEARTH

MASTER
BED RM.
12⁰ x 12²

REF.

PANTRY

LIVING RM.
24⁰ x 15⁶

BEAMED CEILING

CL

CL

BATH

WASH
RM.

LAUNDRY

LINEN

WASH   DRY

BATH

ENTRY

DN.

CURB

PORCH

STUDY-
BED RM.
12⁰ x 11⁶

CL

BED RM.
12⁰ x 11⁶

1540 SQ. FT.

GARAGE
21⁴ x 21⁴

RM-2565C
Cost Estimate

Cost at a Glance

Cost per Square Foot: $93.95
Total Cost: $144,683

Cost by Category

		Cost per Square Foot of Living Area		
		Materials	Installation	Total
1. Site Work	Excavation for the basement and footings.		1.21	1.21
2. Foundation	Main house — 12″ wide concrete masonry unit foundation wall on 20″ x 10″ reinforced concrete perimeter footings. Trench footings — 8″ and 12″ wide reinforced concrete. Slabs — 4″ thick steel trowel finished reinforced concrete over compacted gravel.	5.25	7.02	12.27
3. Framing	Exterior walls — 2 x 4 studs, 16″ on center with 1/2″ plywood sheathing. Floor — 2 x 10 joists, 16″ on center with 3/4″ plywood subfloor. Roof — pre-engineered trusses and site cut rafters with 1/2″ plywood sheathing.	9.63	5.84	15.47
4. Exterior Walls	Texture 1-11 and brick veneer siding over 15# felt vapor barrier with R-13 and R-11 insulation. Vinyl clad casement, and fixed windows and sliding glass patio doors.	12.98	7.36	20.34
5. Roofing	Heavyweight three tab asphalt shingles over 30# felt roofing paper. Aluminum gutters, downspouts, drip edge and flashings.	1.88	1.61	3.49
6. Interiors	Walls and ceilings — 1/2″ and 5/8″ taped and finished gypsum wallboard, primed and painted with one coat latex. Pine interior trim with one coat paint or stain. Flooring — 74% carpet, 16% vinyl, 7% hardwood and 3% ceramic tile.	7.71	6.39	14.10
7. Specialties	Hardwood faced particle board case kitchen cabinets and bathroom vanities with plastic laminate countertops. Washer, dryer, range with hood, single wall oven, dishwasher and refrigerator. One masonry fireplace.	6.55	2.51	9.06
8. Mechanical	Oil fired forced hot air heat with central air conditioning. One full bath, one 1/2 bath and one 3/4 bath. Stainless steel kitchen sink with disposal.	4.72	3.24	7.96
9. Electrical	200 amp service, branch circuit wiring with romex cable. Exterior and interior lighting fixtures, receptacles and switches.	2.45	1.46	3.91
10. Overhead	Contractor's overhead and profit.	3.58	2.56	6.14
Total Cost per Square Foot		$54.75	$39.20	$93.95

To purchase a full set of Sepias, Bill of Materials and Detailed Costs — turn to page 267.

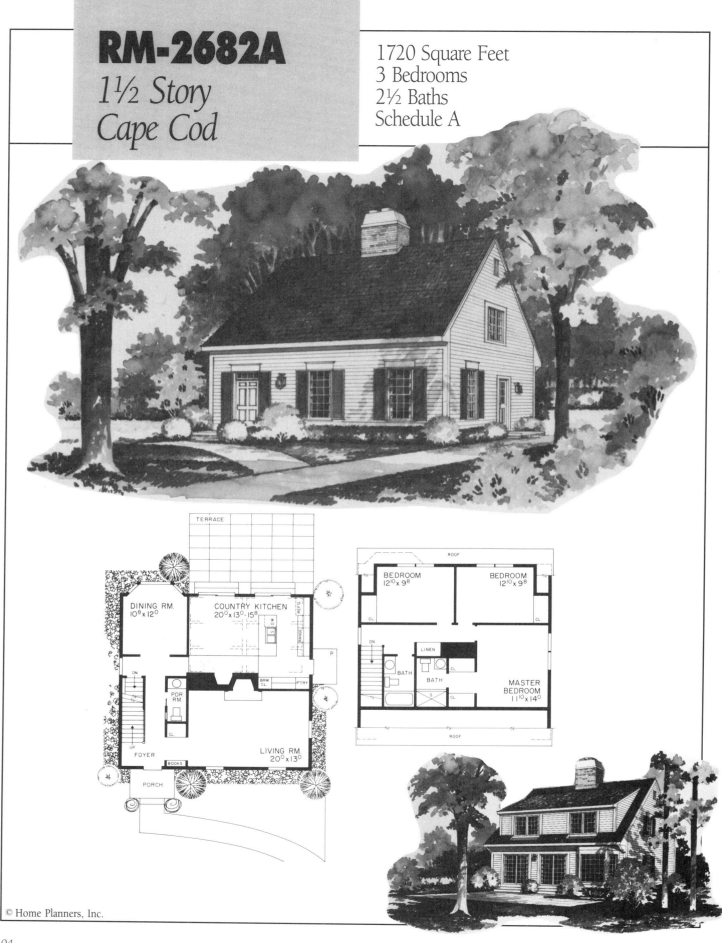

RM-2682A
1½ Story
Cape Cod

1720 Square Feet
3 Bedrooms
2½ Baths
Schedule A

TERRACE

DINING RM.
10⁸ x 12⁰

COUNTRY KITCHEN
20⁰ x 13⁰-15⁸

REF'G
RANGE
DW

P

PDR
RM

BRM
CL

PTRY

DN

CL

UP
FOYER

BOOKS

LIVING RM.
20⁰ x 13⁰

PORCH

ROOF

BEDROOM
12¹⁰ x 9⁸

BEDROOM
12¹⁰ x 9⁸

CL

CL

DN

LINEN

BATH

CL

BATH

CL

S

MASTER
BEDROOM
11¹⁰ x 14⁰

ROOF

RM-2682A
Cost Estimate

Cost by Category		Cost per Square Foot of Living Area		
		Materials	Installation	Total
1. Site Work	Excavation for the basement and footings.		.60	.60
2. Foundation	Main house — 12″ wide masonry foundation wall on 20″ x 10″ reinforced concrete perimeter footings. Trench footings — 8″ and 12″ wide reinforced concrete. Slabs — 4″ thick reinforced steel trowel finished concrete over compacted gravel.	3.44	4.51	7.95
3. Framing	Exterior walls — 2 x 6 studs, 16″ on center with 1/2″ plywood sheathing. Floors — 2 x 10 floor joists, 16″ on center with 3/4″ plywood subfloor. Roof — site cut rafters with 1/2″ plywood sheathing.	7.99	4.90	12.89
4. Exterior Walls	Beveled cedar siding over 15# felt vapor barrier with R-19 wall insulation. Vinyl clad double hung and fixed windows and sliding glass patio doors.	9.01	3.02	12.03
5. Roofing	Heavyweight three tab asphalt shingles over 30# felt roofing paper. Aluminum gutters, downspouts, drip edge and flashings.	.97	.88	1.85
6. Interiors	Walls and ceilings — 1/2″ and 5/8″ taped and finished gypsum wallboard, primed and painted with one coat latex. Pine interior trim, with one coat paint or stain. Flooring — 75% carpet, 13% vinyl, 8% hardwood and 4% ceramic tile.	6.36	5.55	11.91
7. Specialties	Hardwood faced particle board case kitchen cabinets and bathroom vanities with plastic laminate countertops. Washer, dryer, range with hood, dishwasher, and refrigerator. Two masonry fireplaces.	6.41	3.39	9.80
8. Mechanical	Oil fired forced hot air heat with central air conditioning. One full bath, one 3/4 bath and one 1/2 bath. Stainless steel double bowl kitchen sink with disposal.	4.45	3.06	7.51
9. Electrical	200 amp service, branch circuit wiring with romex cable. Exterior and interior lighting fixtures, receptacles and switches.	2.11	1.29	3.40
10. Overhead	Contractor's overhead and profit.	2.85	1.91	4.76
Total Cost per Square Foot		$43.59	$29.11	$72.70

To purchase a full set of Sepias, Bill of Materials and Detailed Costs — turn to page 267.

RM-3571

*1½ Story
Cape Cod*

1747 Square Feet
3 Bedrooms
2½ Baths
Schedule B

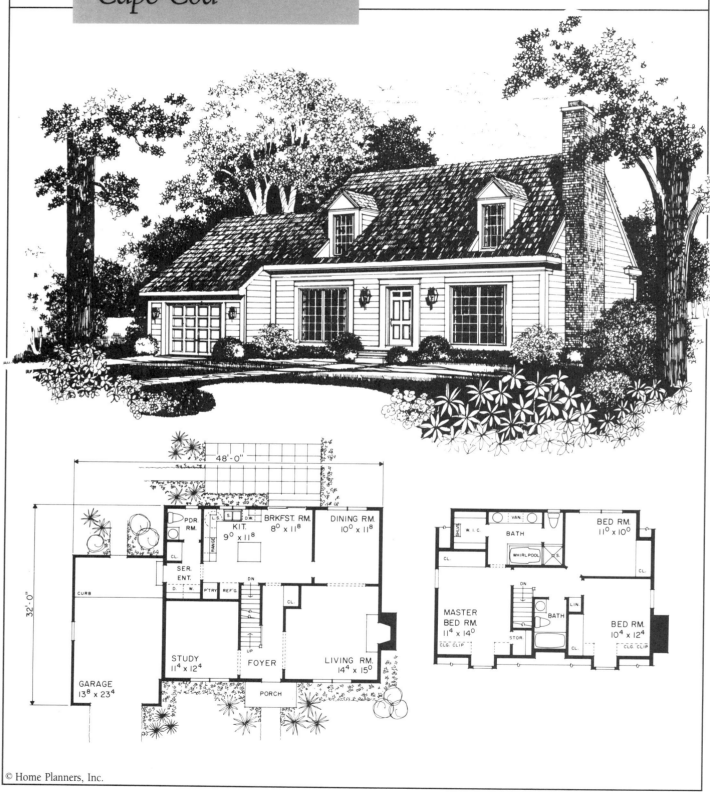

© Home Planners, Inc.

RM-3571
Cost Estimate

Cost at a Glance

Cost per Square Foot: $80.39
Total Cost: $140,441

Cost by Category

		Cost per Square Foot of Living Area		
		Materials	Installation	Total
1. Site Work	Excavation for the basement and footings.		.84	.84
2. Foundation	Main house — 10″ wide reinforced concrete foundation wall on 20″ x 10″ reinforced concrete perimeter footings. Trench footings — 10″ wide reinforced concrete. Slabs — 4″ thick steel trowel finished reinforced concrete over compacted gravel.	3.76	4.68	8.44
3. Framing	Exterior walls — 2 x 6 studs, 16″ on center with 1/2″ plywood sheathing. Garage — 2 x 4 studs, 16″ on center. Roof — pre-engineered trusses and site cut rafters with 1/2″ plywood sheathing.	9.31	5.52	14.83
4. Exterior Walls	Beveled cedar siding over 15# felt vapor barrier with R-19 and R-11 insulation. Vinyl clad double hung and casement windows and a sliding glass patio door.	9.69	4.53	14.22
5. Roofing	Heavyweight three tab asphalt shingles over 30# felt roofing paper. Aluminum gutters, downspouts, drip edge and flashings.	1.33	1.16	2.49
6. Interiors	Walls and ceilings — 1/2″ and 5/8″ taped and finished gypsum wallboard, primed and painted with one coat latex. Pine interior trim with one coat paint or stain. Flooring — 73% carpet, 18% vinyl, 5% hardwood and 4% ceramic tile.	6.94	6.32	13.26
7. Specialties	Hardwood faced particle board case kitchen cabinets and bathroom vanities with plastic laminate countertops. Washer, dryer, range with hood, dishwasher and refrigerator. One masonry fireplace.	5.81	2.40	8.21
8. Mechanical	Oil fired forced hot air heat with central air conditioning. One full bath, one 1/2 bath and a master suite with a whirlpool and shower. Stainless steel kitchen sink with disposal.	5.49	3.39	8.88
9. Electrical	200 amp service, branch circuit wiring with romex cable. Exterior and interior lighting fixtures, receptacles and switches.	2.58	1.38	3.96
10. Overhead	Contractor's overhead and profit.	3.14	2.12	5.26
Total Cost per Square Foot		$48.05	$32.34	$80.39

To purchase a full set of Sepias, Bill of Materials and Detailed Costs — turn to page 267.

RM-2661
1½ Story Cape Cod

1797 Square Feet
3 Bedrooms
2½ Baths
Schedule A

34'-0"

30'-0"

DINING RM.
10⁰ x 13⁶

COUNTRY KITCHEN
23⁰ x 13⁶ + BAY

SEAT

P

RAISED HEARTH

PDR RM

LIVING RM.
13⁰ x 15⁶

FOYER

STUDY
10⁰ x 9⁶

PORCH

ROOF

BATH

BATH

BEDROOM
12⁴ x 11⁰

LINEN

MASTER BEDROOM
13⁰ x 15⁸

WALK IN CLOSET

CEILING CLIP

BEDROOM
11⁰ x 12⁰

ROOF

98

RM-2661
Cost Estimate

Cost at a Glance

Cost per Square Foot: $72.75
Total Cost: $130,731

Cost by Category

		Cost per Square Foot of Living Area		
		Materials	Installation	Total
1. Site Work	Excavation for the basement and footings.		.59	.59
2. Foundation	20″ wide x 10″ thick concrete footing, full height 12″ thick concrete block foundation wall. All slabs are 4″ thick reinforced concrete, steel trowel finish over 4″ compacted gravel base.	2.92	3.71	6.63
3. Framing	Exterior walls — 2 x 4 studs located 16″ on center with 1/2″ plywood sheathing. Floors — 2 x 12 joists, 16″ on center with 3/4″ plywood subfloor. Roof — site cut 2 x 8 rafters spaced 16″ on center.	7.59	4.59	12.18
4. Exterior Walls	Red cedar horizontal beveled wood siding with a 15# felt vapor barrier and 3½″ insulation. Vinyl clad double hung windows with shutters, raised panel entry doors.	7.41	2.63	10.04
5. Roofing	Heavyweight fiberglass roof shingles over 30# felt paper, aluminum flashings, drip edge, gutters and downspouts.	1.19	.95	2.14
6. Interiors	Walls and ceilings — taped and finished gypsum wallboard, primed and painted with one coat latex. Pine door, window, ceiling and baseboard moldings with one coat paint or stain. Carpet — 70%, vinyl — 20%, ceramic tile — 5%, hardwood — 5%.	7.80	6.66	14.46
7. Specialties	Kitchen cabinets and bathroom vanities — hardwood faced particleboard cases with plastic laminate countertops. Three medicine cabinets, masonry fireplace, washer, dryer, oven, cooktop, refrigerator and dishwasher.	7.07	4.03	11.10
8. Mechanical	Oil fired forced hot air heating, central air conditioning, one full bath, one 3/4 bath and one 1/2 bath. Double bowl kitchen sink, disposal.	4.38	2.96	7.34
9. Electrical	200 amp electrical service, branch circuit wiring — romex cable, interior and exterior lighting fixtures, receptacles and switches.	2.25	1.26	3.51
10. Overhead	Contractor's overhead and profit.	2.84	1.92	4.76
	Total Cost per Square Foot	**$43.45**	**$29.30**	**$72.75**

To purchase a full set of Sepias, Bill of Materials and Detailed Costs — turn to page 267.

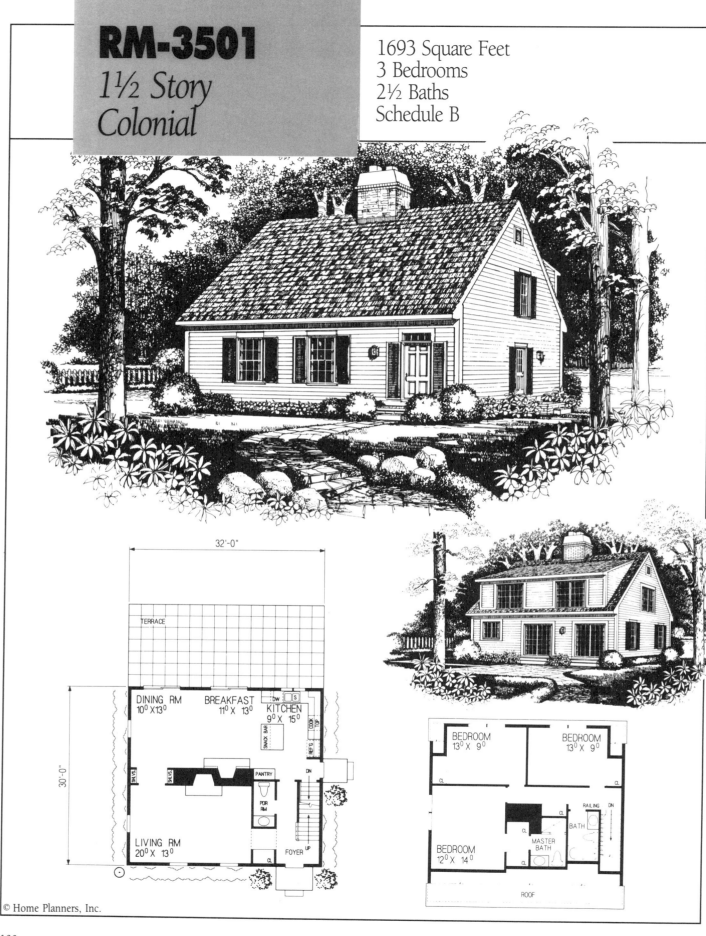

RM-3501
1½ Story Colonial

1693 Square Feet
3 Bedrooms
2½ Baths
Schedule B

32'-0"

TERRACE

DINING RM
10⁰ X 13⁰

BREAKFAST
11⁰ X 13⁰

KITCHEN
9⁰ X 15⁰

DW S
COOK TOP

SNACK BAR

REF'G

30'-0"

SHLVS. SHLVS.

PANTRY

DN

PDR RM

LIVING RM
20⁰ X 13⁰

FOYER UP

BEDROOM
13⁰ X 9⁰

BEDROOM
13⁰ X 9⁰

CL CL

RAILING DN

BATH

MASTER BATH

BEDROOM
12⁰ X 14⁰

CL

ROOF

RM-3501
Cost Estimate

Cost per Square Foot: $73.19
Total Cost: $123,910

Cost by Category

		Cost per Square Foot of Living Area		
		Materials	Installation	Total
1. Site Work	Excavation for the basement and footings.		.60	.60
2. Foundation	Main house — 10″ wide reinforced concrete foundation walls on 20″ x 10″ reinforced concrete perimeter footings. Trench footings — 8″ wide reinforced concrete. Slabs — 4″ thick steel trowel finished reinforced concrete over compacted gravel.	3.20	3.94	7.14
3. Framing	Exterior walls — 2 x 6 studs, 16″ on center with 1/2″ plywood sheathing. Floor — 2 x 10 and 2 x 12 joists, 16″ on center with 3/4″ plywood subfloor. Roof — site cut rafters with 1/2″ plywood sheathing.	8.65	4.97	13.62
4. Exterior Walls	Beveled cedar siding over 15# felt vapor barrier with R-19 insulation. Vinyl clad double hung and casement windows with sliding glass patio doors.	8.98	3.05	12.03
5. Roofing	Heavyweight three tab asphalt shingles over 30# felt roofing paper. Aluminum gutters, downspouts, drip edge and flashings.	.89	.82	1.71
6. Interiors	Walls and ceilings — 1/2″ and 5/8″ taped and finished gypsum wallboard, primed and painted with one coat latex. Pine interior trim with one coat paint or stain. Flooring — 84% carpet, 8% vinyl, 4% hardwood and 4% ceramic tile.	6.46	5.63	12.09
7. Specialties	Hardwood faced particle board case kitchen cabinets and bathroom vanities with plastic laminate countertops. Washer, dryer, range with hood, dishwasher and refrigerator. Two masonry fireplaces.	6.62	3.76	10.38
8. Mechanical	Oil fired forced hot air heat with central air conditioning. One full bath, one 3/4 bath and one 1/2 bath. Stainless steel kitchen sink with disposal.	4.48	3.05	7.53
9. Electrical	200 amp service, branch circuit wiring with romex cable. Exterior and interior lighting fixtures, receptacles and switches.	2.04	1.26	3.30
10. Overhead	Contractor's overhead and profit.	2.89	1.90	4.79
Total Cost per Square Foot		$44.21	$28.98	$73.19

To purchase a full set of Sepias, Bill of Materials and Detailed Costs — *turn to page 267.*

RM-3444

1½ Story Transitional

1973 Square Feet
3 Bedrooms
2½ Baths
Schedule B

RM-3444
Cost Estimate

Cost at a Glance

Cost per Square Foot: $94.46
Total Cost: $186,369

Cost by Category

		Cost per Square Foot of Living Area		
		Materials	Installation	Total
1. Site Work	Excavation for the basement and footings.		.91	.91
2. Foundation	Main house and garage — 10″ wide reinforced concrete foundation wall on 20″ x 10″ reinforced concrete perimeter footings. Trench footings - 8″ wide reinforced concrete. Slabs — 4″ thick steel trowel finished reinforced concrete over compacted gravel.	3.67	4.60	8.27
3. Framing	Exterior walls — 2 x 6 studs, 16″ on center with 1/2″ plywood sheathing. Floors — 2 x 10 joists, 16″ on center with 3/4″ plywood subfloor. Roof — pre-engineered trusses and site cut rafters with 5/8″ plywood sheathing.	11.67	9.28	20.95
4. Exterior Walls	Beveled cedar siding over 15# felt vapor barrier with R-19 and R-11 insulation. Vinyl clad fixed and casement windows. Raised panel front entry door with sidelight and transom window.	11.91	4.16	16.07
5. Roofing	Red cedar shingles 30# felt roofing paper. Aluminum gutters, downspouts, drip edge and flashings.	3.16	1.59	4.75
6. Interiors	Walls and ceilings — 1/2″ and 5/8″ taped and finished gypsum wallboard, primed and painted with one coat latex. Pine interior trim with one coat paint or stain. Flooring — 70% carpet, and 30% vinyl.	7.18	7.10	14.28
7. Specialties	Hardwood faced particle board case kitchen cabinets and bathroom vanities with plastic laminate countertops. Washer, dryer, range with hood, dishwasher and refrigerator. Pre-manufactured fireplace.	6.45	1.68	8.13
8. Mechanical	Oil fired forced hot air heat with central air conditioning. One full bath, one 1/2 bath and a master suite with a whirlpool and shower. Stainless steel kitchen sink with disposal.	6.68	4.02	10.70
9. Electrical	200 amp service, branch circuit wiring with romex cable. Exterior and interior lighting fixtures, receptacles and switches.	2.65	1.57	4.22
10. Overhead	Contractor's overhead and profit.	3.74	2.44	6.18
	Total Cost per Square Foot	**$57.11**	**$37.35**	**$94.46**

To purchase a full set of Sepias, Bill of Materials and Detailed Costs — turn to page 267.

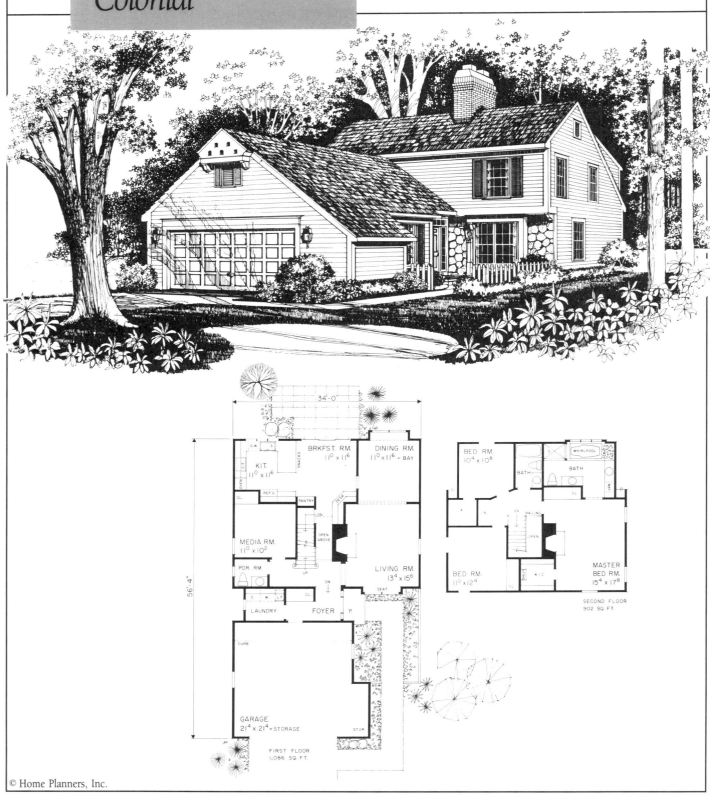

RM-3379

2 Story
Colonial

1988 Square Feet
3 Bedrooms
2½ Baths
Schedule B

34'-0"

56'-4"

KIT.
11⁰ x 11⁶

BRKFST. RM.
11⁰ x 11⁶

DINING RM.
11⁰ x 11⁶ + BAY

MEDIA RM.
11⁰ x 10²

PDR. RM.

LIVING RM.
13⁴ x 15⁶

LAUNDRY   FOYER

GARAGE
21⁴ x 21⁴ + STORAGE

STOR.

FIRST FLOOR
1,086 SQ. FT.

BED. RM.
10⁴ x 10⁸

WHIRLPOOL

BATH   BATH

BED. RM.
11⁰ x 12⁴

MASTER
BED. RM.
15⁴ x 17⁸

SECOND FLOOR
902 SQ. FT.

RM-3379
Cost Estimate

Cost at a Glance

Cost per Square Foot: $81.07
Total Cost: $161,167

Cost by Category		Cost per Square Foot of Living Area		
		Materials	Installation	Total
1. Site Work	Excavation for the basement and footings.		.78	.78
2. Foundation	Main house — 10″ wide reinforced concrete foundation wall on 20″ x 10″ reinforced concrete perimeter footings. Trench footings — 10″ wide reinforced concrete walls. Slabs — 4″ thick steel trowel finished reinforced concrete over compacted gravel.	4.46	4.41	8.87
3. Framing	Exterior walls — 2 x 6 studs, 16″ on center with 1/2″ plywood sheathing. Garage — 2 x 4 studs, 16″ on center. Floor — 2 x 10 and 2 x 12 joists, 16″ on center with 3/4″ plywood subfloor. Roof — pre-engineered trusses and site cut rafters with 1/2″ plywood sheathing.	8.97	5.45	14.42
4. Exterior Walls	Beveled cedar siding over 15# felt vapor barrier with R-19 and R-11 insulation. Vinyl clad fixed, double hung and casement windows, and a sliding glass patio door.	10.73	4.10	14.83
5. Roofing	Heavyweight three tab asphalt shingles over 30# felt roofing paper. Aluminum gutters, downspouts, drip edge and flashings.	1.22	1.14	2.36
6. Interiors	Walls and ceilings — 1/2″ and 5/8″ taped and finished gypsum wallboard, primed and painted with one coat latex. Pine interior trim with one coat paint or stain. Flooring — 71% carpet, 16% vinyl, 7% hardwood and 6% ceramic tile.	7.65	6.73	14.38
7. Specialties	Hardwood faced particle board case kitchen cabinets and bathroom vanities with plastic laminate countertops. Washer, dryer, cooktop with hood, double ovens, dishwasher and refrigerator. One masonry fireplace.	8.91	3.64	12.55
8. Mechanical	Oil fired forced hot air heat with central air conditioning. One full bath, one 1/2 bath and a master suite with a whirlpool and shower. Stainless steel kitchen sink with disposal.	2.35	1.47	3.82
9. Electrical	200 amp service, branch circuit wiring with romex cable. Exterior and interior lighting fixtures, receptacles and switches.	2.39	1.36	3.75
10. Overhead	Contractor's overhead and profit.	3.27	2.04	5.31
Total Cost per Square Foot		$49.95	$31.12	$81.07

To purchase a full set of Sepias, Bill of Materials and Detailed Costs — turn to page 267.

RM-2711

2 Story Contemporary

1999 Square Feet
3 Bedrooms
2½ Baths
Schedule B

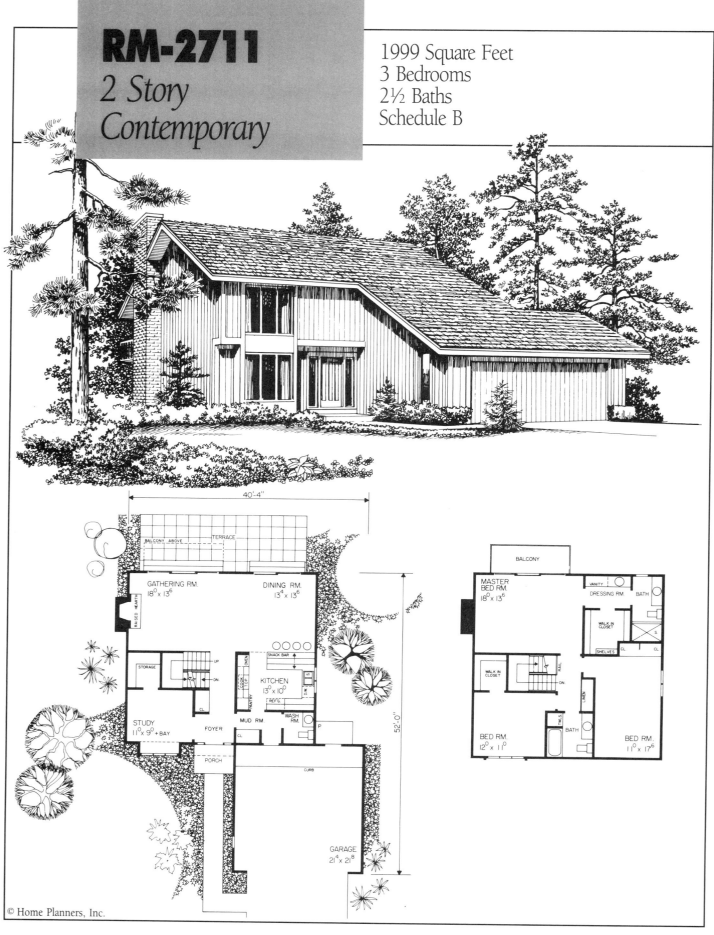

40'-4"

BALCONY ABOVE | TERRACE

GATHERING RM.
18⁰ x 13⁶

DINING RM.
13⁴ x 13⁶

RAISED HEARTH

STORAGE

UP
DN.

CL

OVEN

COOK TOP

PANTRY

SNACK BAR

KITCHEN
13⁰ x 10⁰

D.W.

REFG.

STUDY
11⁰ x 9⁰ + BAY

FOYER

MUD RM.

WASH RM.

CL

P

PORCH

CURB

52'-0"

GARAGE
21⁴ x 21⁸

BALCONY

MASTER BED RM.
18⁰ x 13⁶

VANITY

DRESSING RM.

BATH

WALK IN CLOSET

S

WALK IN CLOSET

SHELVES

CL

CL

UP
DN.

HALL

LINEN

BED RM.
12⁰ x 11⁰

THLS

BATH

BED RM.
11⁰ x 17⁶

RM-2711
Cost Estimate

Cost at a Glance

Cost per Square Foot: $67.63
Total Cost: $135,192

Cost by Category

		Cost per Square Foot of Living Area		
		Materials	Installation	Total
1. Site Work	Excavation for the basement and footings.		.73	.73
2. Foundation	Main house — 12″ concrete block wall on a 20″ x 10″ reinforced concrete footing. Garage — 8″ x 42″ reinforced concrete foundation wall. Slabs — 4″ thick reinforced steel trowel finished concrete on 4″ compacted gravel.	1.74	2.12	3.86
3. Framing	Main house — 2 x 6 and 2 x 4 studs, 16″ on center with 1/2″ plywood sheathing. Garage — 2 x 4 studs, 16″ on center. Floor — 2 x 10 joists, 16″ on center with 3/4″ tongue and groove plywood subfloor. Roof — pre-engineered trusses with 1/2″ plywood sheathing.	7.73	4.85	12.58
4. Exterior Walls	Vertical pine board siding with 15# felt vapor barrier. Vinyl clad casement windows and sliding glass doors, 3 panel entry door. R-19 and R-11 insulation.	8.43	3.42	11.85
5. Roofing	Heavyweight three tab asphalt roof shingles on 30# felt paper. Aluminum drip edge, flashings, gutters and downspouts.	1.31	1.03	2.34
6. Interiors	Wall finish — one coat primer and one coat paint on 1/2″ or 5/8″ gypsum wallboard. Flooring — 82% carpet, 8% vinyl, 6% ceramic tile, 4% hardwood.	6.95	6.13	13.08
7. Specialties	Hardwood faced, particle board case kitchen cabinets and bath vanities with plastic laminate countertops. Washer, dryer, cooktop with hood, double ovens, dishwasher and refrigerator. One masonry fireplace, one cantilevered balcony.	5.68	2.14	7.82
8. Mechanical	Oil fired forced hot air heat with central air conditioning. One full bath, one 3/4 bath and one 1/2 bath. Double bowl kitchen sink with disposal.	4.25	2.93	7.18
9. Electrical	200 amp service, branch circuit wiring with romex cable. Exterior and interior lighting fixtures, receptacles and switches.	2.45	1.31	3.76
10. Overhead	Contractor's overhead and profit.	2.70	1.73	4.43
	Total Cost per Square Foot	**$41.24**	**$26.39**	**$67.63**

To purchase a full set of Sepias, Bill of Materials and Detailed Costs — turn to page 267.

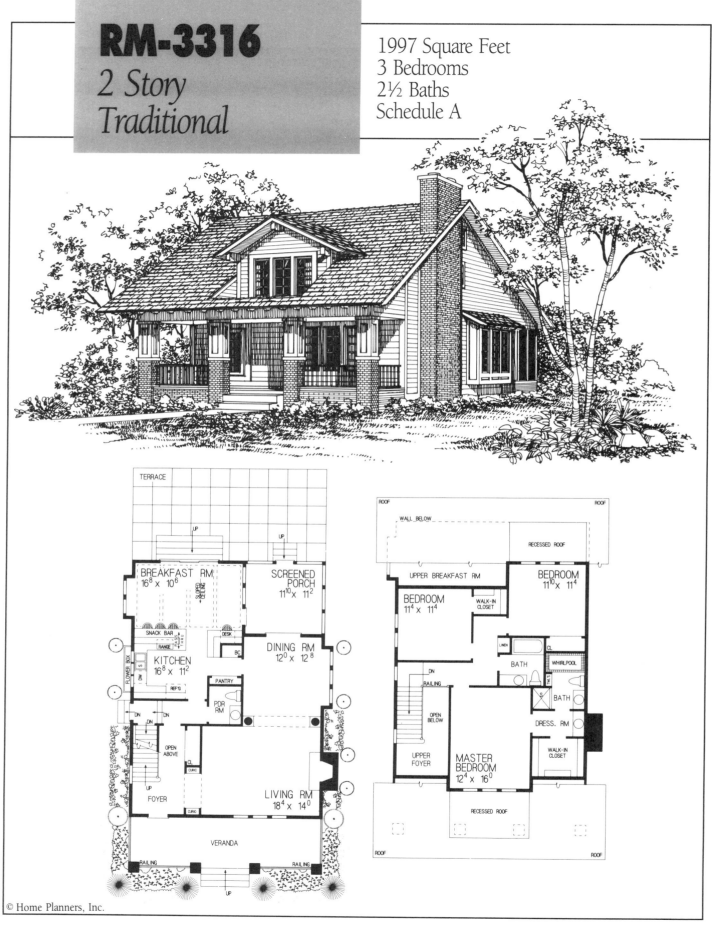

RM-3316

2 Story Traditional

1997 Square Feet
3 Bedrooms
2½ Baths
Schedule A

First floor:
- TERRACE
- UP
- BREAKFAST RM 16⁸ x 10⁶
- SCREENED PORCH 11¹⁰ x 11²
- SLOPED CEILING
- SNACK BAR
- PASS THRU
- DESK
- RANGE
- BC
- DINING RM 12⁰ x 12⁸
- FLOWER BOX
- DW
- S
- KITCHEN 16⁸ x 11²
- REF'G
- PANTRY
- PDR RM
- DN
- DN
- OPEN ABOVE
- CL
- CURIO
- UP
- FOYER
- CURIO
- LIVING RM 18⁴ x 14⁰
- VERANDA
- RAILING
- RAILING
- UP

Second floor:
- ROOF
- ROOF
- WALL BELOW
- RECESSED ROOF
- UPPER BREAKFAST RM
- BEDROOM 11¹⁰ x 11⁴
- BEDROOM 11⁴ x 11⁴
- WALK-IN CLOSET
- LINEN
- CL
- BATH
- WHIRLPOOL
- DN
- RAILING
- TWL S
- BATH
- S
- OPEN BELOW
- DRESS. RM
- UPPER FOYER
- MASTER BEDROOM 12⁴ x 16⁰
- WALK-IN CLOSET
- RECESSED ROOF
- ROOF
- ROOF

RM-3316
Cost Estimate

Cost at a Glance

Cost per Square Foot: $87.41
Total Cost: $174,557

Cost by Category

		Cost per Square Foot of Living Area		
		Materials	Installation	Total
1. Site Work	Excavation for the basement and footings.		.56	.56
2. Foundation	Main house — 10″ wide reinforced concrete foundation wall on 20″ x 10″ reinforced concrete perimeter footings. Slabs — 4″ thick reinforced steel trowel finished concrete over compacted gravel.	3.50	4.26	7.76
3. Framing	Exterior walls — 2 x 6 studs, 16″ on center with 1/2″ plywood sheathing. Floors — 2 x 8 floor joists, 16″ on center with 3/4″ plywood subfloor. Roof — site cut 2 x 6, 2 x 8 and 2 x 12 rafters with 1/2″ plywood sheathing.	11.66	6.48	18.14
4. Exterior Walls	Beveled cedar siding over 15# felt vapor barrier with R-19 wall insulation. Vinyl clad fixed and casement windows and sliding glass doors. Fully glazed front entry door.	14.98	4.85	19.83
5. Roofing	Heavyweight three tab asphalt shingles over 30# felt roofing paper. Aluminum gutters, downspouts, drip edge and flashings.	1.59	1.16	2.75
6. Interiors	Walls and ceilings — 1/2″ and 5/8″ taped and finished gypsum wallboard, primed and painted with one coat latex. Pine interior trim. Flooring — 65% carpet, 22% vinyl, 8% ceramic tile, and 5% hardwood.	6.60	5.98	12.58
7. Specialties	Hardwood faced particle board case kitchen cabinets and bathroom vanities with plastic laminate countertops. Washer, dryer, range with hood, dishwasher, and refrigerator. Masonry fireplace.	5.21	2.30	7.51
8. Mechanical	Oil fired forced hot air heat with central air conditioning. One 1/2 bath, one full bath and a master suite with a whirlpool and shower. Stainless steel double bowl kitchen sink with disposal.	5.58	3.21	8.79
9. Electrical	200 amp service, branch circuit wiring with romex cable. Exterior and interior lighting fixtures, receptacles and switches.	2.45	1.32	3.77
10. Overhead	Contractor's overhead and profit.	3.61	2.11	5.72
Total Cost per Square Foot		$55.18	$32.23	$87.41

To purchase a full set of Sepias, Bill of Materials and Detailed Costs — turn to page 267.

RM-1956A

2 *Story*
Traditional

1718 Square Feet
4 Bedrooms
2½ Baths
Schedule A

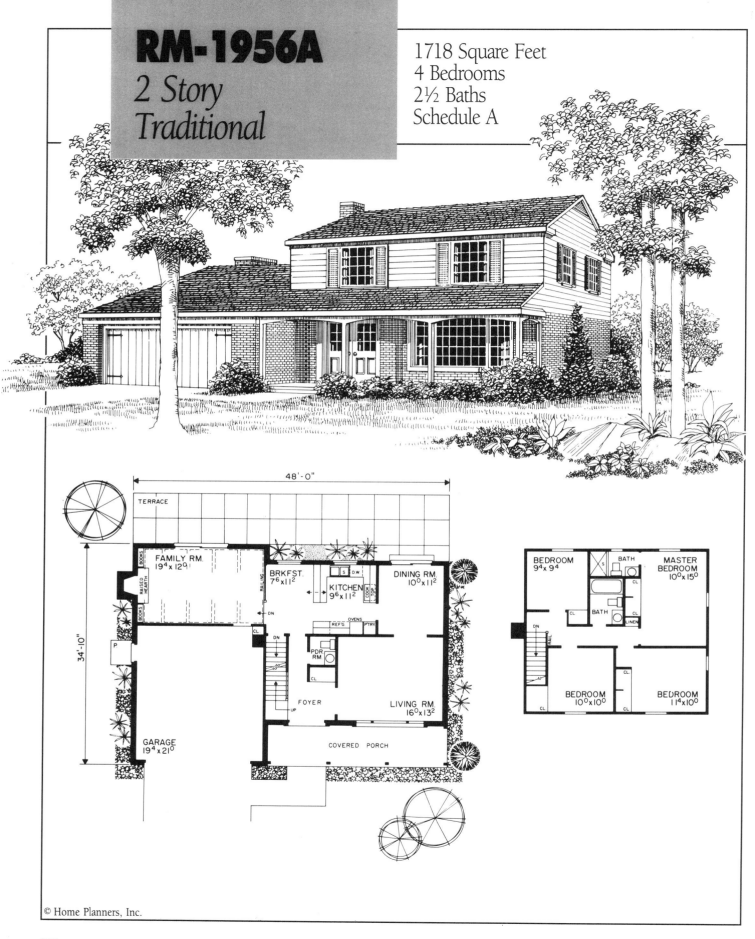

TERRACE

48'-0"

FAMILY RM.
19⁴ x 12⁹

BRKFST.
7⁶ x 11²

KITCHEN
9⁶ x 11²

DINING RM.
10⁰ x 11²

34'-10"

LIVING RM.
16⁰ x 13²

FOYER

GARAGE
19⁴ x 21⁰

COVERED PORCH

BEDROOM
9⁴ x 9⁴

BATH

MASTER BEDROOM
10⁰ x 15⁰

BATH

LINEN

BEDROOM
10⁰ x 10⁰

BEDROOM
11⁴ x 10⁰

RM-1956A
Cost Estimate

Cost at a Glance

Cost per Square Foot: $79.13
Total Cost: $135,945

Cost by Category		Materials	Installation	Total
		Cost per Square Foot of Living Area		
1. Site Work	Excavation for the basement and footings.		.86	.86
2. Foundation	Main house — 12″ wide masonry foundation wall on 20″ x 10″ reinforced concrete perimeter footings. Trench footings — 8″ and 12″ wide reinforced concrete. Slabs — 4″ thick reinforced steel trowel finished concrete over compacted gravel.	3.77	4.81	8.58
3. Framing	Exterior walls — 2 x 4 studs, 16″ on center with 1/2″ plywood sheathing. Garage — 2 x 4 studs, 16″ on center. Floors — 2 x 10 floor joists, 16″ on center with 3/4″ plywood subfloor. Roof — pre-engineered trusses and site cut rafters with 1/2″ plywood sheathing.	7.52	4.85	12.37
4. Exterior Walls	Masonry veneer and beveled cedar siding over 15# felt vapor barrier with R-19 and R-11 wall insulation. Vinyl clad fixed, double hung and casement windows and sliding glass patio doors.	10.81	5.48	16.29
5. Roofing	Heavyweight three tab asphalt shingles over 30# felt roofing paper. Aluminum gutters, downspouts, drip edge and flashings.	1.35	1.19	2.54
6. Interiors	Walls and ceilings — 1/2″ and 5/8″ taped and finished gypsum wallboard, primed and painted with one coat latex. Pine interior trim. Flooring — 70% carpet, 22% vinyl, 4% hardwood and 4% ceramic tile.	7.30	6.28	13.58
7. Specialties	Hardwood faced particle board case kitchen cabinets and bathroom vanities with plastic laminate countertops. Washer, dryer, cooktop with hood, double ovens, dishwasher, and refrigerator.	6.12	2.65	8.77
8. Mechanical	Oil fired forced hot air heat with central air conditioning. One full bath, one 3/4 bath and one 1/2 bath. Stainless steel double bowl kitchen sink with disposal.	4.36	3.01	7.37
9. Electrical	200 amp service, branch circuit wiring with romex cable. Exterior and interior lighting fixtures, receptacles and switches.	2.32	1.27	3.59
10. Overhead	Contractor's overhead and profit.	3.05	2.13	5.18
Total Cost per Square Foot		$46.60	$32.53	$79.13

To purchase a full set of Sepias, Bill of Materials and Detailed Costs — turn to page 267.

111

RM-2488

2 Story Tudor

1656 Square Feet
3 Bedrooms
2 Baths
Schedule A

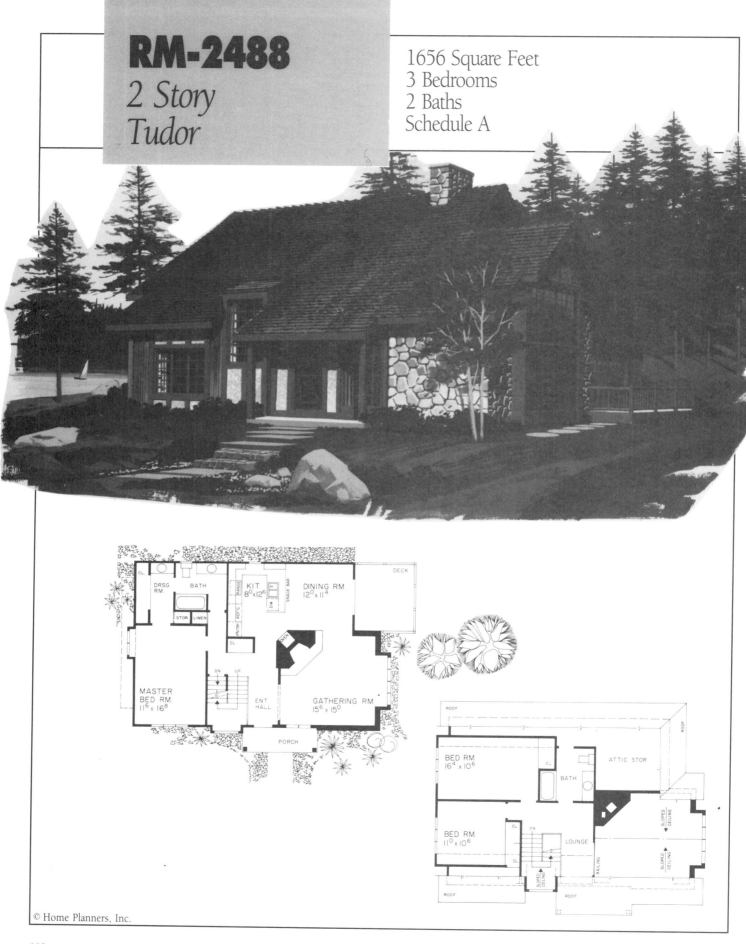

RM-2488
Cost Estimate

Cost at a Glance

Cost per Square Foot: $80.40
Total Cost: $133,142

Cost by Category

		Cost per Square Foot of Living Area		
		Materials	Installation	Total
1. Site Work	Excavation for the basement and footings.		.68	.68
2. Foundation	12″ wide concrete block foundation wall, on 20″ x 10″ poured concrete footings. Basement slab is 4″ thick reinforced concrete, steel trowel finish over 4″ compacted gravel base.	2.77	3.80	6.57
3. Framing	Exterior walls — 2 x 6 and 2 x 4 studs located 16″ on center with 1/2″ plywood sheathing. Floors — 2 x 10 joists, 16″ on center with 3/4″ plywood subfloor. Roof — site cut 2 x 8 rafters 16″ on center with 1/2″ plywood sheathing.	9.20	5.36	14.56
4. Exterior Walls	Combination of pine vertical board siding, fieldstone veneer and stucco with a 15# felt vapor barrier and 3½″ insulation. Vinyl clad casement and fixed glass windows, embossed steel entry door and a 6′ vinyl clad sliding glass door.	12.44	5.72	18.16
5. Roofing	Heavyweight three tab fiberglass roof shingles over 15# felt paper, aluminum flashings and drip edge.	1.73	1.40	3.13
6. Interiors	Walls and ceilings — taped and finished gypsum wallboard primed and painted, one coat latex. Pine door, window and baseboard moldings — one coat paint or stain. Carpet — 85%, vinyl — 1%, ceramic tile — 6%, hardwood — 8%.	6.84	5.97	12.81
7. Specialties	Kitchen cabinets and bathroom vanities — hardwood faced particle board cases, plastic laminate countertops. Two medicine cabinets, masonry fireplace, oven, cooktop, refrigerator and dishwasher.	6.35	2.68	9.03
8. Mechanical	Oil fired forced hot air heating, central air conditioning, 2 full baths. Double bowl kitchen sink, disposal.	4.22	2.77	6.99
9. Electrical	200 amp electrical service, branch circuit wiring--romex cable, interior and exterior lighting fixtures, receptacles and switches.	2.03	1.18	3.21
10. Overhead	Contractor's overhead and profit.	3.19	2.07	5.26
Total Cost per Square Foot		$48.77	$31.63	$80.40

To purchase a full set of Sepias, Bill of Materials and Detailed Costs — turn to page 267.

RM-3331

2 *Story* Tudor

1805 Square Feet
3 Bedrooms
2 Baths
Schedule A

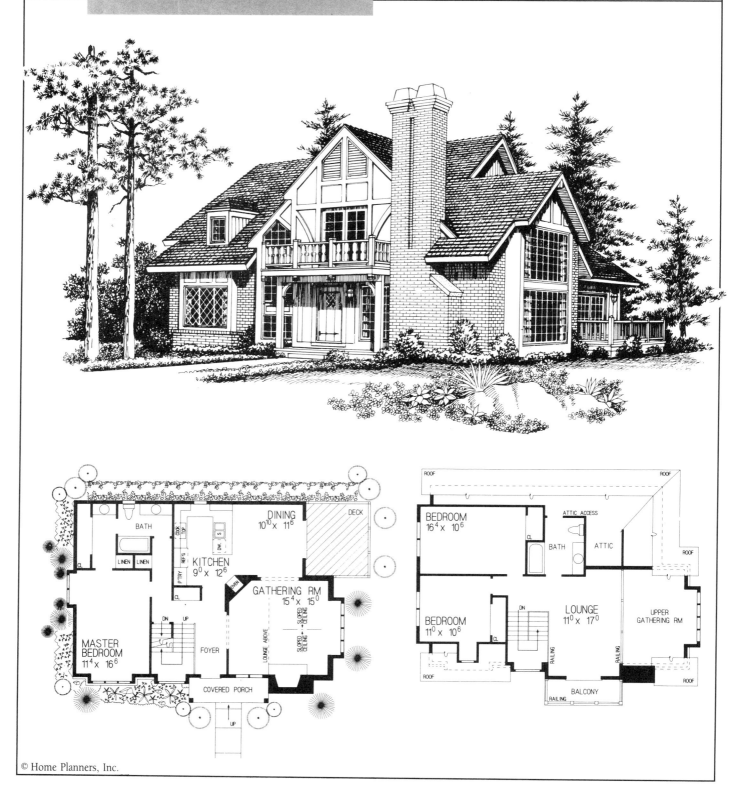

First Floor:

- BATH
- LINEN / LINEN
- CL
- MASTER BEDROOM 11⁴ x 16⁶
- DN / UP
- FOYER
- PTRY
- CL
- KITCHEN 9⁰ x 12⁶
- DINING 10¹⁰ x 11⁶
- DECK
- COOK TOP / REF'G / DW. / S
- OVEN
- GATHERING RM 15⁴ x 15⁰
- LOUNGE ABOVE
- SLOPED CEILING
- COVERED PORCH
- UP

Second Floor:

- ROOF
- BEDROOM 16⁴ x 10⁶
- CL
- ATTIC ACCESS
- BATH
- ATTIC
- ROOF
- BEDROOM 11⁰ x 10⁶
- CL
- DN
- LOUNGE 11⁰ x 17⁰
- UPPER GATHERING RM
- RAILING
- RAILING
- BALCONY
- RAILING
- ROOF

RM-3331
Cost Estimate

Cost by Category

		Cost per Square Foot of Living Area		
		Materials	Installation	Total
1. Site Work	Excavation for the basement and footings.		.63	.63
2. Foundation	Main house — 8″ and 10″ wide reinforced concrete foundation walls on 20″ x 10″ reinforced concrete perimeter footings. Stem walls — 12″ wide reinforced concrete. Slabs — 4″ thick reinforced steel trowel finished concrete over compacted gravel.	2.97	3.50	6.47
3. Framing	Exterior walls — 2 x 6 studs, 16″ on center with 15/32″ plywood sheathing. Floors — 2 x 10 floor joists, 16″ on center with 23/32″ plywood subfloor. Roof — site cut rafters with 15/32″ plywood sheathing.	9.68	5.91	15.59
4. Exterior Walls	Masonry veneer and Texture 1-11 siding over 15# felt vapor barrier with R-19 wall insulation. Vinyl clad fixed, and casement windows and swinging patio doors.	12.74	6.28	19.02
5. Roofing	Wood shingles over 30# felt roofing paper. Aluminum drip edge and flashings.	3.30	1.39	4.69
6. Interiors	Walls and ceilings — 1/2″ and 5/8″ taped and finished gypsum wallboard, primed and painted with one coat latex. Pine interior trim. Flooring — 74% carpet, 14% vinyl, 8% hardwood and 4% ceramic tile.	7.13	6.21	13.34
7. Specialties	Hardwood faced particle board case kitchen cabinets and bathroom vanities with plastic laminate countertops. Washer, dryer, cooktop with hood, double ovens, dishwasher, and refrigerator. One masonry fireplace.	4.78	1.17	5.95
8. Mechanical	Oil fired forced hot air heat with central air conditioning. Two full baths. Stainless steel double bowl kitchen sink with disposal.	3.99	2.63	6.62
9. Electrical	200 amp service, branch circuit wiring with romex cable. Exterior and interior lighting fixtures, receptacles and switches.	2.10	1.21	3.31
10. Overhead	Contractor's overhead and profit.	3.27	2.03	5.30
Total Cost per Square Foot		$49.96	$30.96	$80.92

RM-2974

2 *Story*
Victorian

1772 Square Feet
3 Bedrooms
2½ Baths
Schedule A

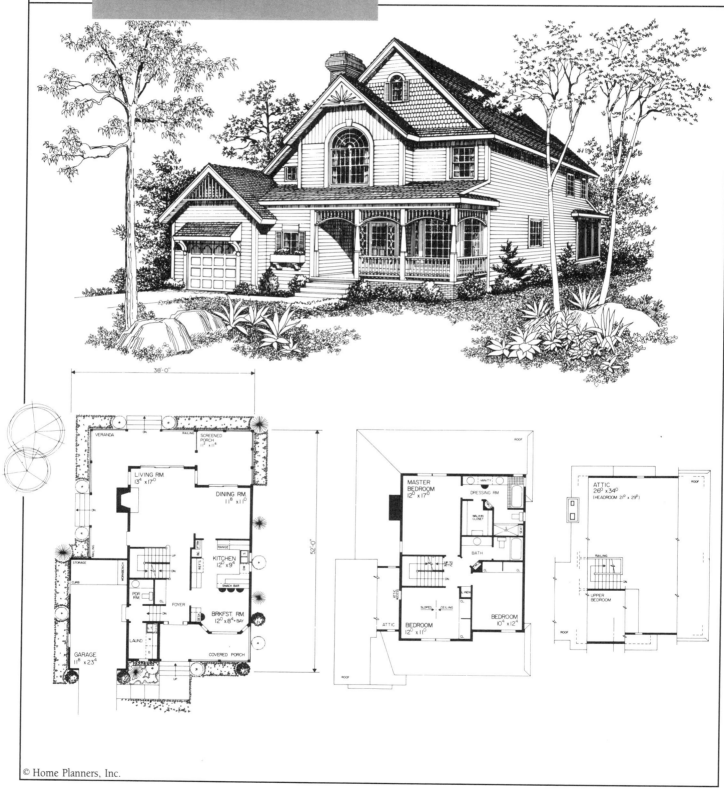

RM-2974
Cost Estimate

Cost by Category

		Cost per Square Foot of Living Area		
		Materials	Installation	Total
1. Site Work	Excavation for the basement and footings.		.80	.80
2. Foundation	Main house—10″ thick reinforced concrete walls on 20″ x 10″ reinforced concrete footings. Garage and porches—10″ x 54″ and 10″ x 66″ reinforced concrete trench walls. Slabs—4″ thick reinforced steel trowel finished concrete on 4″ compacted gravel.	5.22	6.53	11.75
3. Framing	Main house—2 x 6 studs, 16″ on center with 1/2″ plywood sheathing. Garage—2 x 4 studs, 16″ on center. Floor—2 x 10 joists, 16″ on center with 3/4″ tongue and groove plywood subfloor. Roof— pre-engineered roof trusses and site cut 2 x 6 and 2 x 8 rafters, 1/2″ plywood sheathing.	10.73	6.86	17.59
4. Exterior Walls	Horizontal beveled cedar siding, vertical board siding and fancy cut shingles over 15# felt vapor barrier. Vinyl clad double hung, casement and fixed windows and sliding glass doors. Glazed French and paneled entry doors. R-19 and R-11 insulation.	16.61	7.35	23.96
5. Roofing	Heavyweight three tab asphalt roof shingles on 30# felt paper. Aluminum drip edge, flashings, gutters and downspouts.	2.07	1.80	3.87
6. Interiors	Wall finish—1/2″ or 5/8″ gypsum wallboard with one coat primer and one coat finish paint. Pine door, window and baseboard moldings with one coat paint or stain. Flooring—70% carpet, 16% vinyl, 8% ceramic tile and 6% hardwood.	10.36	7.64	18.00
7. Specialties	Hardwood faced, particle board case kitchen cabinets and bath vanities with plastic laminate countertops. Washer, dryer, range with hood, double ovens, dishwasher and refrigerator. A masonry fireplace, covered and screened porches.	5.92	2.27	8.19
8. Mechanical	Oil fired forced hot air heat with central air conditioning. One full bath, one 1/2 bath and a master suite with a whirlpool and shower. Double bowl kitchen sink with disposal.	5.48	3.46	8.94
9. Electrical	200 amp service, branch circuit wiring with romex cable. Exterior and interior lighting fixtures, receptacles and switches.	2.63	1.43	4.06
10. Overhead	Contractor's overhead and profit.	4.13	2.67	6.80
	Total Cost per Square Foot	$63.15	$40.81	$103.96

To purchase a full set of Sepias, Bill of Materials and Detailed Costs—turn to page 267.

RM-3385

2 Story
Victorian

1996 Square Feet
4 Bedrooms
2½ Baths
Schedule C

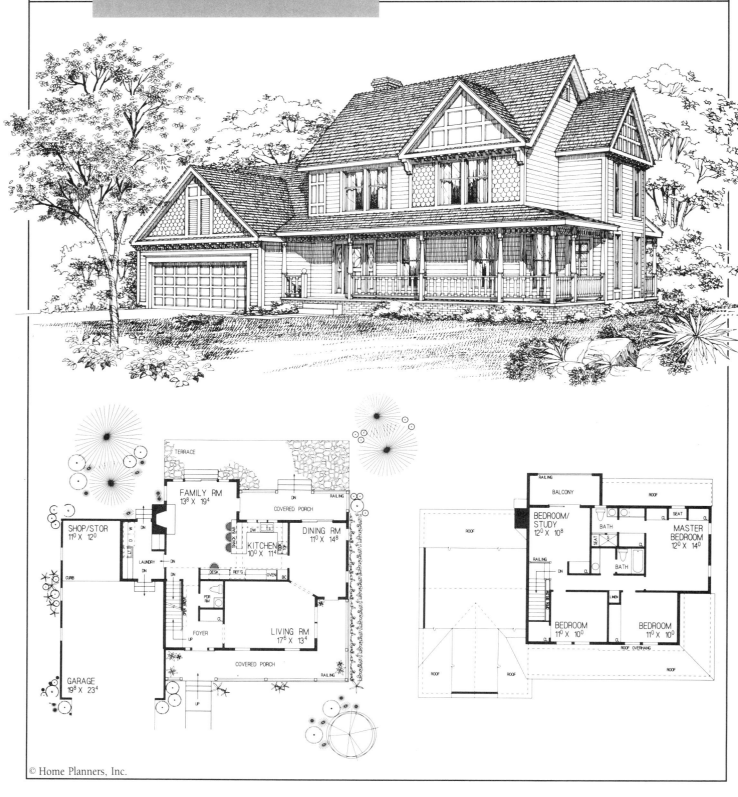

TERRACE

FAMILY RM
13⁸ X 19⁴

COVERED PORCH

RAILING

SHOP/STOR
11⁰ X 12⁰

DINING RM
11⁰ X 14⁸

KITCHEN
10⁰ X 11⁴

SNACK BAR

DW

LAUNDRY

DESK REF'G OVEN BC

PDR RM

CL

FOYER
UP

LIVING RM
17⁶ X 13⁴

CURB

GARAGE
19⁸ X 23⁴

COVERED PORCH

RAILING

RAILING

BALCONY

ROOF

BEDROOM/STUDY
12⁰ X 10⁸

BATH

SEAT

MASTER BEDROOM
12⁰ X 14⁰

ROOF

RAILING

DN

BATH

CL

LINEN

BEDROOM
11⁰ X 10⁰

BEDROOM
11⁰ X 10⁰

ROOF OVERHANG

ROOF

ROOF

ROOF

RM-3385
Cost Estimate

Cost by Category

		Cost per Square Foot of Living Area		
		Materials	Installation	Total
1. **Site Work**	Excavation for the basement and footings.		.78	.78
2. **Foundation**	Main house — 12″ wide masonry foundation wall on 20″ x 10″ reinforced concrete perimeter footings. Trench footings — 8″ and 12″ wide reinforced concrete. Slabs — 4″ thick reinforced steel trowel finished concrete over compacted gravel.	4.14	5.38	9.52
3. **Framing**	Exterior walls — 2 x 6 studs, 16″ on center with 1/2″ plywood sheathing. Garage — 2 x 4 studs, 16″ on center. Floors — 2 x 10 floor joists, 16″ on center with 3/4″ plywood subfloor. Roof — pre-engineered trusses and site cut rafters with 1/2″ plywood sheathing.	9.78	6.19	15.97
4. **Exterior Walls**	Fancy cut shingles and beveled cedar siding over 15# felt vapor barrier with R-19 and R-11 wall insulation. Vinyl clad fixed, double hung and casement windows and sliding glass patio doors.	20.03	9.02	29.05
5. **Roofing**	Heavyweight three tab asphalt shingles over 30# felt roofing paper. Aluminum gutters, downspouts, drip edge and flashings.	2.15	1.65	3.80
6. **Interiors**	Walls and ceilings — 1/2″ and 5/8″ taped and finished gypsum wallboard, primed and painted with one coat latex. Pine interior trim, with one coat paint or stain. Flooring — 80% carpet, 11% vinyl, 4% hardwood and 5% ceramic tile.	7.10	6.45	13.55
7. **Specialties**	Hardwood faced particle board case kitchen cabinets and bathroom vanities with plastic laminate countertops. Washer, dryer, cooktop with hood, double ovens, dishwasher, and refrigerator. One masonry fireplace.	7.20	3.05	10.25
8. **Mechanical**	Oil fired forced hot air heat with central air conditioning. One full bath, one 3/4 bath and one 1/2 bath. Double bowl stainless steel kitchen sink with disposal.	4.25	2.91	7.16
9. **Electrical**	200 amp service, branch circuit wiring with romex cable. Exterior and interior lighting fixtures, receptacles and switches.	2.52	1.34	3.86
10. **Overhead**	Contractor's overhead and profit.	4.00	2.57	6.57
	Total Cost per Square Foot	**$61.17**	**$39.34**	**$100.51**

To purchase a full set of Sepias, Bill of Materials and Detailed Costs — turn to page 267.

RM-2608
Multilevel Traditional

1912 Square Feet
4 Bedrooms
2½ Baths
Schedule A

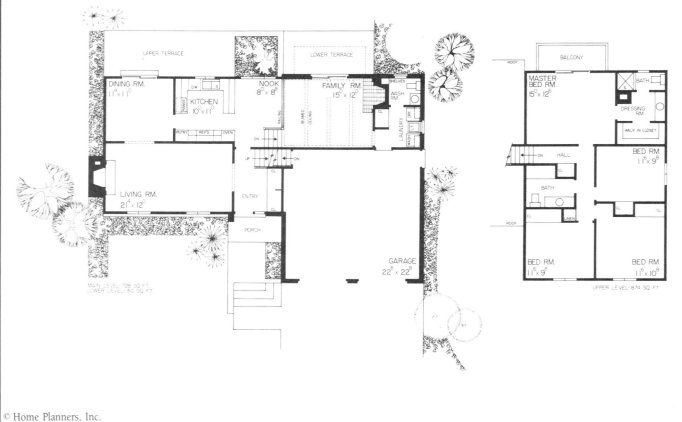

UPPER TERRACE

DINING RM.
11⁴ x 11⁰

KITCHEN
10⁰ x 11⁰

NOOK
8⁰ x 8⁸

LOWER TERRACE

FAMILY RM.
15⁰ x 12⁰

SHELVES

WASH RM.

LAUNDRY

D.W.

RANGE

PANTRY   REFS.   OVEN

DN

UP   DN

LIVING RM.
21⁴ x 12⁰

ENTRY

PORCH

GARAGE
22⁸ x 22⁸

MAIN LEVEL 729 SQ. FT.
LOWER LEVEL 310 SQ. FT.

ROOF   BALCONY

MASTER BED RM.
15⁰ x 12⁶

BATH

DRESSING RM.

WALK IN CLOSET

DN   HALL

BED RM.
11⁶ x 9⁸

BATH

ROOF   CL   LINEN

BED RM.
11⁶ x 9⁸

BED RM.
11⁶ x 10⁸

UPPER LEVEL 874 SQ. FT.

RM-2608

Cost Estimate

Cost at a Glance

Cost per Square Foot: $79.30
Total Cost: $151,621

Cost by Category

		Cost per Square Foot of Living Area		
		Materials	Installation	Total
1. Site Work	Excavation for the basement and footings.		.67	.67
2. Foundation	Main house — 12″ wide concrete masonry unit foundation wall on 20″ x 10″ reinforced concrete perimeter footings. Trench footings — 8″ and 12″ wide reinforced concrete. Slabs — 4″ thick reinforced steel trowel finished concrete over compacted gravel.	4.15	5.36	9.51
3. Framing	Exterior walls — 2 x 6 studs, 16″ on center with 1/2″ plywood sheathing. Garage — 2 x 4 studs, 16″ on center. Floors — 2 x 8 floor joists, 16″ on center with 3/4″ plywood subfloor. Roof — pre-engineered trusses and site cut rafters with 1/2″ plywood sheathing.	7.65	4.91	12.56
4. Exterior Walls	Masonry veneer and Texture 1-11 siding over 15# felt vapor barrier with R-19 and R-11 wall insulation. Vinyl clad casement windows and sliding patio doors.	9.69	4.58	14.27
5. Roofing	Red cedar shake shingles over 30# felt roofing paper. Aluminum gutters, downspouts, and flashings.	2.29	1.73	4.02
6. Interiors	Walls and ceilings — 1/2″ and 5/8″ taped and finished gypsum wallboard, primed and painted with one coat latex. Pine interior trim, with one coat paint or stain. Flooring — 80% carpet, 15% vinyl, 3% hardwood and 2% ceramic tile.	7.02	6.08	13.10
7. Specialties	Hardwood faced particle board case kitchen cabinets and bathroom vanities with plastic laminate countertops. Washer, dryer, cooktop with hood, double ovens, dishwasher, and refrigerator. Two masonry fireplaces.	6.09	2.77	8.86
8. Mechanical	Oil fired forced hot air heat with central air conditioning. One full bath, one 3/4 bath and one 1/2 bath. Stainless steel double bowl kitchen sink with disposal.	4.26	2.87	7.13
9. Electrical	200 amp service, branch circuit wiring with romex cable. Exterior and interior lighting fixtures, receptacles and switches.	2.58	1.41	3.99
10. Overhead	Contractor's overhead and profit.	3.06	2.13	5.19
	Total Cost per Square Foot	$46.79	$32.51	$79.30

To purchase a full set of Sepias, Bill of Materials and Detailed Costs — turn to page 267.

2000 to 2500 Square Feet

Plans	Style	Stories	Total SF	Bedrms	Baths	Page
RM3496	Ranch	1 Story	2033	3	2	124
RM3332	Ranch	1 Story	2168	3	2½	126
RM3421	Southwestern	1 Story	2145	3	2½	128
RM3601	Traditional	1 Story	2424	3	2½	130
RM3560	Transitional	1 Story	2189	3	2	132
RM3346	Tudor	1 Story	2032	3	2	134
RM2657	Cape Cod	1½ Story	2085	3	2½	136
RM3372	Cape Cod	1½ Story	2201	3	2½	138
RM2563	Cape Cod	1½ Story	2190	4	2	140
RM2927	Contemporary	1½ Story	2129	3	2	142
RM3476	Contemporary	1½ Story	2008	3	2½	144
RM3455	Contemporary	1½ Story	2075	3	2½	146
RM3338	Contemporary	1½ Story	2284	3	2½	148
RM3458	Contemporary	1½ Story	2343	4	2½	150
RM3461	Farmhouse	1½ Story	2002	4	2½	152
RM3318A	Traditional	1½ Story	2097	2	2	154
RM2733	Colonial	2 Story	2180	4	2½	156
RM3484	Contemporary	2 Story	2087	3	2½	158
RM3477	Contemporary	2 Story	2152	3	2½	160
RM2776	Farmhouse	2 Story	2008	3	2½	162
RM2774	Farmhouse	2 Story	2339	4	2½	164
RM3564	Traditional	2 Story	2041	3	2½	166
RM3562	Transitional	2 Story	2109	3	2½	168
RM2854	Tudor	2 Story	2211	3	2½	170
RM3309	Victorian	2 Story	2391	3	2½	172
RM1850	Traditional	Multilevel	2184	3	3	174

RM-3496

1 Story Ranch

2033 Square Feet
3 Bedrooms
2 Baths
Schedule B

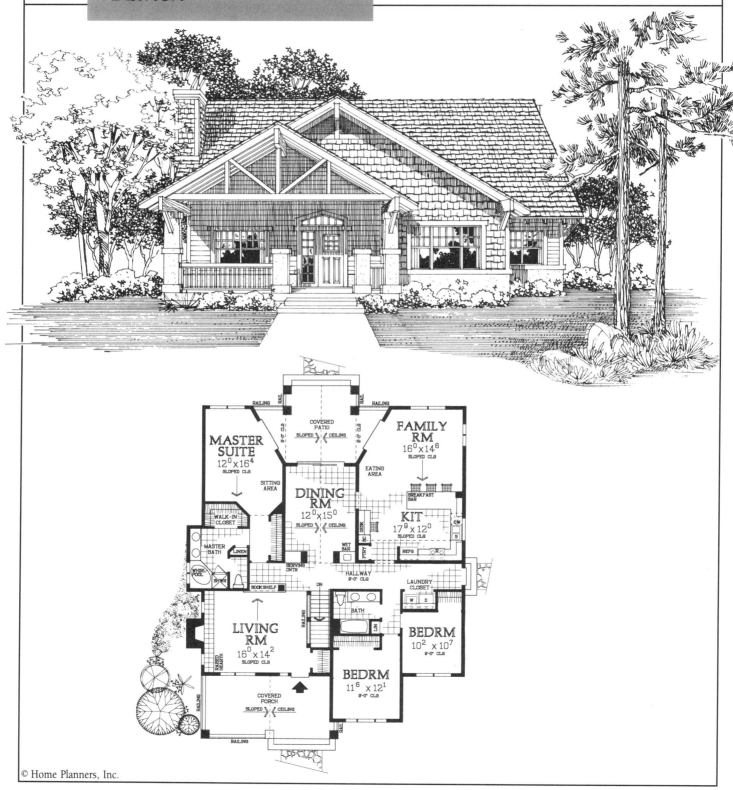

MASTER SUITE
12⁰ x 16⁴
SLOPED CLG

SITTING AREA

WALK-IN CLOSET

MASTER BATH

LINEN

WHIRL-POOL

SHWR

BOOK SHELF

RAISED HEARTH

LIVING RM
16⁰ x 14²
SLOPED CLG

RAILING

DN

COVERED PORCH
SLOPED CEILING

RAILING

COVERED PATIO
SLOPED CEILING

RAILING

DINING RM
12⁰ x 15⁰
SLOPED CEILING

WET BAR

SERVING CNTR

HALLWAY

BATH

LIN

BEDRM
11⁶ x 12¹
8'-0" CLG

EATING AREA

FAMILY RM
16⁰ x 14⁶
SLOPED CLG

BREAKFAST BAR

KIT
17⁹ x 12⁰
SLOPED CLG

REFG

PTRY

LAUNDRY CLOSET

W   D

BEDRM
10² x 10⁷
8'-0" CLG

RM-3496
Cost Estimate

Cost at a Glance
Cost per Square Foot: $93.36
Total Cost: $189,800

Cost by Category

No. Category	Description	Materials	Installation	Total
1. Site Work	Excavation for the basement and footings.		.86	.86
2. Foundation	Main house — 10″ wide reinforced concrete foundation walls on 20″ x 10″ reinforced concrete perimeter footings. Trench footings — 8″ wide reinforced concrete. Slabs — 4″ thick steel trowel finished reinforced concrete over compacted gravel.	5.89	6.78	12.67
3. Framing	Exterior walls — 2 x 6 studs, 16″ on center with 1/2″ plywood sheathing. Floor — 2 x 10 joists, 16″ on center with 3/4″ plywood subfloor. Roof — pre-engineered trusses and site cut rafters with 5/8″ plywood sheathing.	12.89	8.70	21.59
4. Exterior Walls	Beveled cedar horizontal and cedar shake siding over 15# felt vapor barrier with R-19 insulation. Vinyl clad fixed, sliding and double hung windows with sliding and swinging glass patio doors.	9.82	3.45	13.27
5. Roofing	Red cedar shingle roofing over 30# felt roofing paper. Aluminum gutters, downspouts, drip edge and flashings.	3.26	1.51	4.77
6. Interiors	Walls and ceilings — 1/2″ and 5/8″ taped and finished gypsum wallboard, primed and painted with one coat latex. Pine interior trim with one coat paint or stain. Flooring — 58% carpet, and 42% ceramic tile.	6.86	6.41	13.27
7. Specialties	Hardwood faced particle board case kitchen cabinets and bathroom vanities with plastic laminate countertops. Washer, dryer, range with hood, 2 microwave ovens, dishwasher and refrigerator. One pre-manufactured fireplace.	6.75	1.68	8.43
8. Mechanical	Oil fired forced hot air heat with central air conditioning. One full bath, and a master suite with a tub and shower. Stainless steel kitchen sink with disposal.	5.01	3.08	8.09
9. Electrical	200 amp service, branch circuit wiring with romex cable. Exterior and interior lighting fixtures, receptacles and switches.	2.76	1.54	4.30
10. Overhead	Contractor's overhead and profit.	3.73	2.38	6.11
Total Cost per Square Foot		**$56.97**	**$36.39**	**$93.36**

Cost per Square Foot of Living Area

To purchase a full set of Sepias, Bill of Materials and Detailed Costs — turn to page 267.

RM-3332

1 Story Ranch

2168 Square Feet
3 Bedrooms
2½ Baths
Schedule B

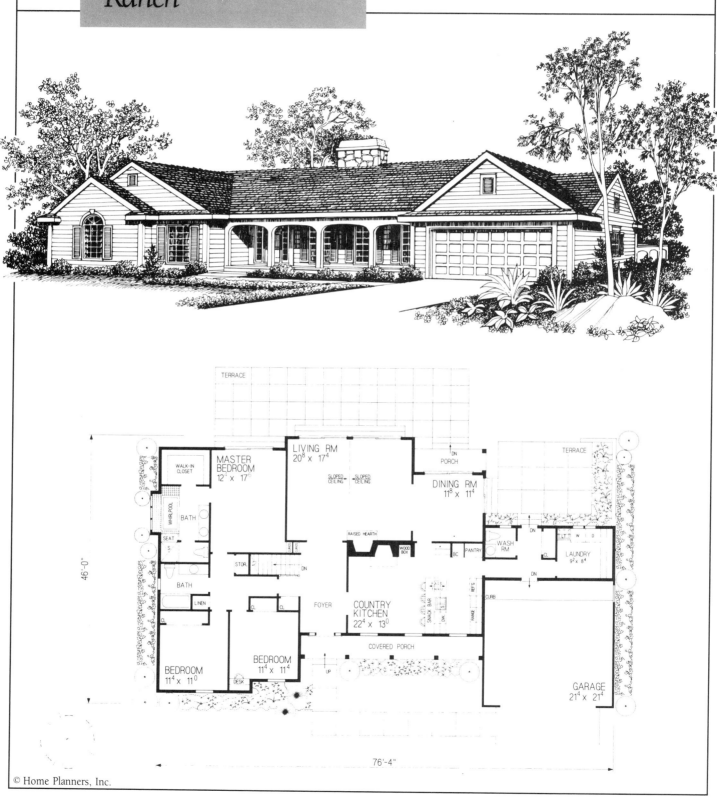

RM-3332
Cost Estimate

Cost at a Glance

Cost per Square Foot: $89.52
Total Cost: $194,079

Cost by Category

		Cost per Square Foot of Living Area		
		Materials	Installation	Total
1. Site Work	Excavation for the basement and footings.		1.05	1.05
2. Foundation	Main house — 10″ thick reinforced concrete walls on 20″ wide x 10″ high reinforced concrete footings. Garage and porches — 8″ wide x 42″ and 51″ high reinforced concrete trench walls. Slabs — 4″ thick reinforced steel trowel finished concrete on 4″ compacted gravel.	5.41	6.77	12.18
3. Framing	Main house — 2 x 6 studs, 16″ on center with 1/2″ plywood sheathing. Garage — 2 x 4 studs, 16″ on center. Floor — 2 x 10 joists, 16″ on center with 3/4″ tongue and groove plywood subfloor. Roof — pre-engineered roof trusses and site cut rafters with 1/2″ plywood sheathing.	10.40	5.97	16.37
4. Exterior Walls	Beveled cedar siding over 15# felt vapor barrier. Vinyl clad double hung, casement and fixed windows and sliding glass doors. Paneled and crossbuck entry doors. R-19 and R-11 insulation.	11.79	3.40	15.19
5. Roofing	Heavyweight three tab asphalt roof shingles on 30# felt paper. Aluminum drip edge, flashing, gutters and downspouts.	2.52	1.85	4.37
6. Interiors	Walls and ceilings — 1/2″ or 5/8″ gypsum, primed and painted one coat. Pine door, window and baseboard moldings. Flooring — 60% carpet, 25% vinyl, 8% ceramic tile and 7% hardwood.	6.96	5.97	12.93
7. Specialties	Kitchen cabinets and bathroom vanities — hardwood faced, with plastic laminate countertops. Washer, dryer, range with hood, dishwasher and refrigerator. Two masonry fireplaces, a covered porch and terrace.	7.38	2.58	9.96
8. Mechanical	Oil fired forced hot air heat with central air conditioning. One full bath, one 1/2 bath and a master suite with a whirlpool tub and shower. Double bowl kitchen sink with disposal.	5.29	3.03	8.32
9. Electrical	200 amp service, branch circuit wiring with romex cable. Exterior and interior lighting fixtures, receptacles and switches.	2.14	1.16	3.30
10. Overhead	Contractor's overhead and profit.	3.63	2.22	5.85
	Total Cost per Square Foot	$55.52	$34.00	$89.52

To purchase a full set of Sepias, Bill of Materials and Detailed Costs — turn to page 267.

RM-3421

1 Story Southwestern

2145 Square Feet
3 Bedrooms
2½ Baths
Schedule B

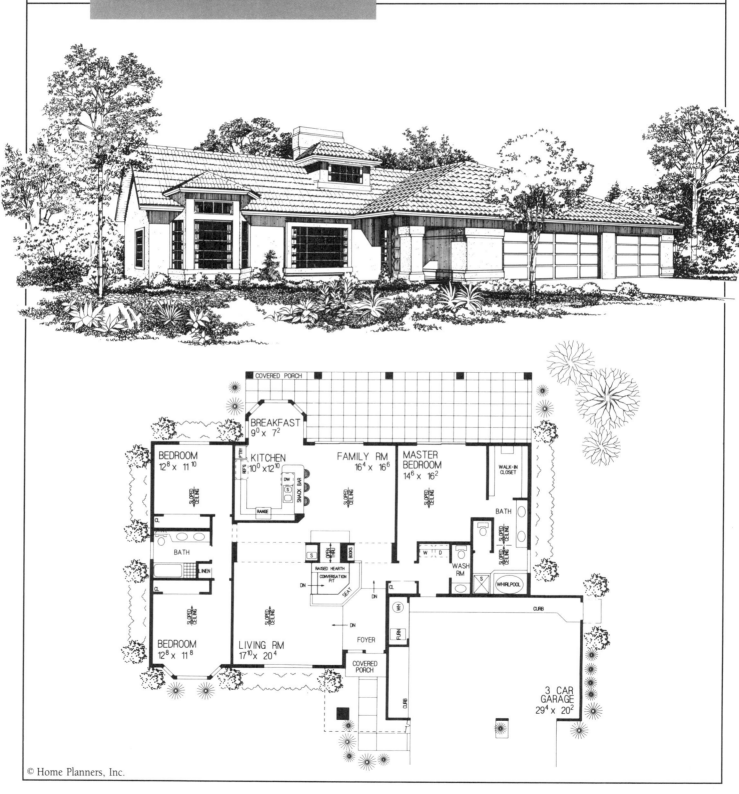

COVERED PORCH

BREAKFAST
9⁰ x 7²

BEDROOM
12⁸ x 11¹⁰

KITCHEN
10⁰ x 12¹⁰

FAMILY RM
16⁴ x 16⁶

MASTER BEDROOM
14⁶ x 16²

WALK-IN CLOSET

SNACK BAR

RANGE

DW

BATH

LINEN

RAISED HEARTH

CONVERSATION PIT

SEAT

BATH

WASH RM

WHIRLPOOL

BEDROOM
12⁸ x 11⁸

LIVING RM
17¹⁰ x 20⁴

FOYER

COVERED PORCH

WH

FURN

CURB

3 CAR GARAGE
29⁴ x 20²

© Home Planners, Inc.

128

RM-3421
Cost Estimate

Cost at a Glance

Cost per Square Foot: $81.18
Total Cost: $174,131

Cost by Category

		Cost per Square Foot of Living Area		
		Materials	Installation	Total
1. Site Work	Excavation for the slab and footings.		.34	.34
2. Foundation	Main house—6″ x 24″ reinforced concrete walls on 6″ x 16″ reinforced concrete footings. Spread footings and continuous footings at interior bearing points. Slabs—4″ thick reinforced steel trowel finished concrete on 4″ compacted gravel.	2.61	3.25	5.86
3. Framing	Main house—2 x 6 studs, 16″ on center with 1/2″ plywood sheathing. Garage—2 x 4 studs, 16″ on center. Roof—pre-engineered roof trusses and 2 x 6 and 2 x 8 site cut roof rafters with 5/8″ plywood sheathing.	9.59	10.53	20.12
4. Exterior Walls	Three coat stucco on high ribbed metal lath siding over 15# felt vapor barrier. Vinyl clad casement and fixed windows and sliding glass doors. Paneled front entry door and flush mechanical room door. R-19 and R-11 insulation.	6.88	1.99	8.87
5. Roofing	Concrete mission barrel type tile roofing over 30# felt roofing paper. Aluminum drip edge, flashings, gutters and downspouts.	8.98	2.82	11.80
6. Interiors	Wall finish—1/2″ or 5/8″ gypsum wallboard with one coat primer and one coat finish paint. Pine door, window and baseboard moldings. Flooring—76% carpet, 15% vinyl, 6% hardwood and 3% ceramic tile.	5.91	5.38	11.29
7. Specialties	Hardwood faced, particle board case kitchen cabinets and bath vanities with plastic laminate countertops. Washer, dryer, range with hood, dishwasher and refrigerator. A pre-fabricated twin faced fireplace and a covered patio.	4.76	1.19	5.95
8. Mechanical	Oil fired forced hot air heat with central air conditioning. One full bath, one 1/2 bath and a master suite with a whirlpool and shower. Double bowl kitchen sink with disposal.	4.96	3.06	8.02
9. Electrical	200 amp service, branch circuit wiring with romex cable. Exterior and interior lighting fixtures, receptacles and switches.	2.41	1.21	3.62
10. Overhead	Contractor's overhead and profit.	3.23	2.08	5.31
Total Cost per Square Foot		$49.33	$31.85	$81.18

To purchase a full set of Sepias, Bill of Materials and Detailed Costs—turn to page 267.

RM-3601

1 Story
Traditional

2424 Square Feet
3 Bedrooms
2½ Baths
Schedule C

© Home Planners, Inc.

RM-3601
Cost Estimate

Cost at a Glance

Cost per Square Foot: $98.49
Total Cost: $238,739

Cost by Category

		Cost per Square Foot of Living Area		
		Materials	Installation	Total
1. Site Work	Excavation for the basement and footings.		1.00	1.00
2. Foundation	Main house—10″ and 24″ wide reinforced concrete foundation walls on 20″ x 10″ and 24″ x 10″ reinforced concrete perimeter footings. Trench footings—8″ wide reinforced concrete. Slabs—4″ thick reinforced steel trowel finished concrete over compacted gravel.	5.09	6.51	11.60
3. Framing	Exterior walls—2 x 6 studs, 16″ on center with 1/2″ plywood sheathing. Floors—2 x 10 floor joists, 16″ on center with 3/4″ plywood subfloor. Roof—pre-engineered trusses and site cut rafters with 5/8″ plywood sheathing.	12.52	8.53	21.05
4. Exterior Walls	Masonry veneer and beveled cedar siding over 15# felt vapor barrier with R-19 and R-11 wall insulation. Vinyl clad fixed, and double hung windows and sliding patio doors.	14.16	7.13	21.29
5. Roofing	Heavyweight three tab asphalt shingles over 30# felt roofing paper. Aluminum gutters, downspouts, drip edge and flashings.	1.97	1.62	3.59
6. Interiors	Walls and ceilings—1/2″ and 5/8″ taped and finished gypsum wallboard, primed and painted with one coat latex. Pine interior trim, with one coat paint or stain. Flooring—63% carpet, and 37% ceramic tile.	7.06	6.43	13.49
7. Specialties	Hardwood faced particle board case kitchen cabinets and bathroom vanities with plastic laminate countertops. Washer, dryer, cooktop with hood, double ovens, dishwasher, and refrigerator. One masonry fireplace.	5.93	2.64	8.57
8. Mechanical	Oil fired forced hot air heat with central air conditioning. One full bath and one 1/2 bath and a master suite with a whirlpool and shower. Stainless steel double bowl kitchen sink with disposal.	4.58	2.88	7.46
9. Electrical	200 amp service, branch circuit wiring with romex cable. Exterior and interior lighting fixtures, receptacles and switches.	2.51	1.48	3.99
10. Overhead	Contractor's overhead and profit.	3.77	2.68	6.45
	Total Cost per Square Foot	$57.59	$40.90	$98.49

To purchase a full set of Sepias, Bill of Materials and Detailed Costs—turn to page 267.

RM-3560

1 Story Transitional

2189 Square Feet
3 Bedrooms
2 Baths
Schedule B

56'-0"

72'-0"

BED RM.
13⁸ x 12⁰

LIVING RM.
18⁴ x 20⁰

DINING RM.
9⁴ x 13⁰

MASTER BED RM.
15⁰ x 18⁰

SLOPED CEILING

HIS W.I.C.

HER W.I.C.

LEDGE ABOVE

BATH

DESK

COOK TOP

SLOPED CEILING

WHIRLPOOL

FOYER

KITCHEN
19⁴ x 17⁸

OVEN

SER. ENT.

STOR.

PORCH

MEDIA/BED RM.
13⁸ x 15⁸

CURB

GARAGE
22⁰ x 21⁸

RM-3560
Cost Estimate

Cost at a Glance

Cost per Square Foot: $75.09
Total Cost: $164,372

Cost by Category

Category	Description	Materials	Installation	Total
		Cost per Square Foot of Living Area		
1. Site Work	Excavation for the slab and frost walls.		.33	.33
2. Foundation	Main house—12″ wide x 42″ high reinforced concrete frost walls. Slabs—4″ thick steel trowel finished reinforced concrete over compacted gravel.	3.61	4.72	8.33
3. Framing	Exterior walls—2 x 6 studs, 16″ on center with 1/2″ plywood sheathing. Garage—2 x 4 studs, 16″ on center. Roof—pre-engineered trusses and site cut rafters with 1/2″ plywood sheathing.	6.22	4.84	11.06
4. Exterior Walls	Exterior insulation finish system (1½″ thick) siding over 15# felt vapor barrier with R-19 and R-11 insulation. Vinyl clad fixed and casement windows with swinging and sliding glass doors.	8.33	2.84	11.17
5. Roofing	Heavyweight three tab asphalt shingles over 30# felt roofing paper. Aluminum gutters, downspouts, drip edge and flashings.	2.03	1.81	3.84
6. Interiors	Walls and ceilings—1/2″ and 5/8″ taped and finished gypsum wallboard, primed and painted with one coat latex. Pine interior trim with one coat paint or stain. Flooring—66% carpet, 25% vinyl, 5% hardwood and 4% ceramic tile.	6.43	5.69	12.12
7. Specialties	Hardwood faced particle board case kitchen cabinets and bathroom vanities with plastic laminate countertops. Washer, dryer, cooktop with hood, double ovens, dishwasher and refrigerator. One pre-fabricated fireplace.	10.27	1.76	12.03
8. Mechanical	Oil fired forced hot air heat with central air conditioning. One full bath and a master suite with a whirlpool and shower. Stainless steel kitchen sink with disposal.	4.93	3.06	7.99
9. Electrical	200 amp service, branch circuit wiring with romex cable. Exterior and interior lighting fixtures, receptacles and switches.	2.13	1.17	3.30
10. Overhead	Contractor's overhead and profit.	3.08	1.84	4.92
	Total Cost per Square Foot	**$47.03**	**$28.06**	**$75.09**

To purchase a full set of Sepias, Bill of Materials and Detailed Costs—turn to page 267.

RM-3346

1 Story Tudor

2032 Square Feet
2 Bedrooms
2 Baths
Schedule B

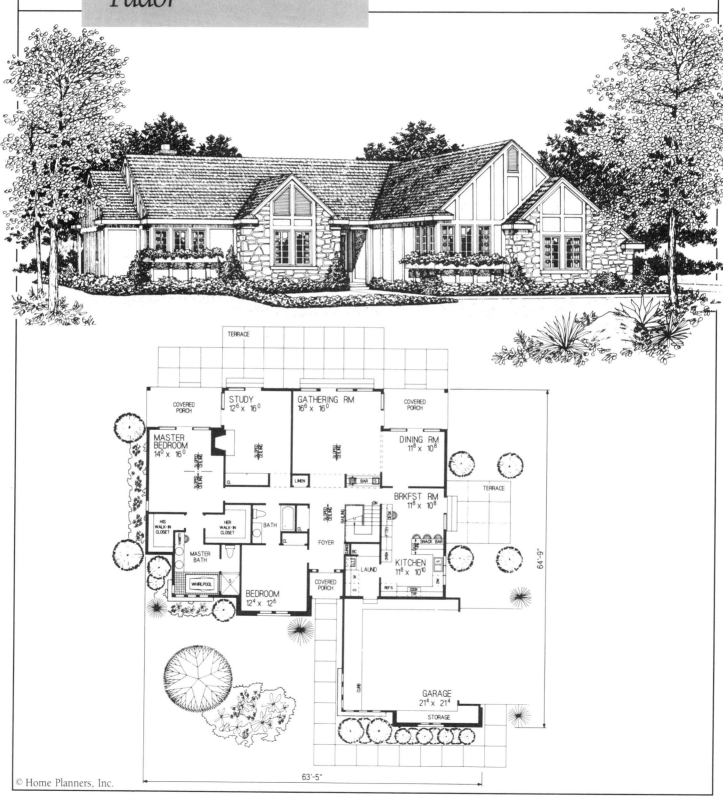

TERRACE

COVERED PORCH

STUDY
12⁶ x 16⁰

GATHERING RM
16⁶ x 16⁰

COVERED PORCH

MASTER BEDROOM
14⁰ x 16⁰

DINING RM
11⁸ x 10⁸

LINEN

BRKFST RM
11⁸ x 10⁸

TERRACE

HIS WALK-IN CLOSET

HER WALK-IN CLOSET

BATH

VANITY

FOYER

DESK

SNACK BAR

MASTER BATH

WHIRLPOOL

BEDROOM
12⁴ x 12⁶

COVERED PORCH

LAUND

KITCHEN
11⁸ x 10¹⁰

REF'G

COOK TOP

DW

GARAGE
21⁴ x 21⁴

STORAGE

64'-9"

63'-5"

RM-3346
Cost Estimate

Cost at a Glance

Cost per Square Foot: $93.01
Total Cost: $188,996

Cost by Category

		Cost per Square Foot of Living Area		
		Materials	Installation	Total
1. Site Work	Excavation for the basement and footings.		1.07	1.07
2. Foundation	Main house — 10″ wide reinforced concrete foundation wall on 20″ x 10″ reinforced concrete perimeter footings. Trench footings — 8″ wide reinforced concrete. Slabs — 4″ thick steel trowel finished reinforced concrete over compacted gravel.	5.61	7.00	12.61
3. Framing	Exterior walls — 2 x 6 studs, 16″ on center with 1/2″ plywood sheathing. Garage — 2 x 4 studs, 16″ on center. Floor — 2 x 10 joists, 16″ on center with 3/4″ plywood subfloor. Roof — site cut rafters with 1/2″ plywood sheathing.	11.77	6.65	18.42
4. Exterior Walls	Stone veneer and 1″ stucco system siding over 15# felt vapor barrier with R-19 and R-11 insulation. Vinyl clad fixed and casement windows and swinging glass patio doors.	13.35	4.42	17.77
5. Roofing	Heavyweight three tab asphalt shingles over 30# felt roofing paper. Aluminum gutters, downspouts, drip edge and flashings.	1.96	1.88	3.84
6. Interiors	Walls and ceilings — 1/2″ and 5/8″ taped and finished gypsum wallboard, primed and painted with one coat latex. Pine interior trim. Flooring — 69% carpet, 19% vinyl, 4% hardwood and 8% ceramic tile.	6.70	6.48	13.18
7. Specialties	Hardwood faced particle board case kitchen cabinets and bathroom vanities with plastic laminate countertops. Washer, dryer, cooktop with hood, double wall ovens, dishwasher and refrigerator. One masonry fireplace.	5.82	2.34	8.16
8. Mechanical	Oil fired forced hot air heat with central air conditioning. One full bath and a master suite with a whirlpool and shower. Stainless steel kitchen sink with disposal.	5.47	3.07	8.54
9. Electrical	200 amp service, branch circuit wiring with romex cable. Exterior and interior lighting fixtures, receptacles and switches.	2.19	1.15	3.34
10. Overhead	Contractor's overhead and profit.	3.70	2.38	6.08
	Total Cost per Square Foot	**$56.57**	**$36.44**	**$93.01**

To purchase a full set of Sepias, Bill of Materials and Detailed Costs — turn to page 267.

RM-2657

*1½ Story
Cape Cod*

2085 Square Feet
3 Bedrooms
2½ Baths
Schedule B

© Home Planners, Inc.

RM-2657
Cost Estimate

Cost at a Glance
Cost per Square Foot: $81.16
Total Cost: $169,218

Cost by Category

		Cost per Square Foot of Living Area		
		Materials	Installation	Total
1. Site Work	Excavation for the basement and footings.		.79	.79
2. Foundation	Main house — 12″ wide concrete masonry unit foundation wall on 20″ x 10″ reinforced concrete perimeter footings. Trench footings — 8″ wide reinforced concrete. Slabs — 4″ thick steel trowel finished reinforced concrete over compacted gravel.	3.62	4.79	8.41
3. Framing	Exterior walls — 2 x 6 studs, 16″ on center with 1/2″ plywood sheathing. Garage — 2 x 4 studs, 16″ on center. Floor — 2 x 10 and 2 x 12 joists, 16″ on center with 3/4″ plywood subfloor. Roof — site cut rafters with 1/2″ plywood sheathing.	9.33	5.67	15.00
4. Exterior Walls	Beveled cedar siding over 15# felt vapor barrier with R-19 and R-11 insulation. Vinyl clad fixed, casement, and double hung windows and swinging glass patio doors.	12.07	4.00	16.07
5. Roofing	Heavyweight three tab asphalt shingles over 30# felt roofing paper. Aluminum gutters, downspouts, drip edge and flashings.	1.38	1.19	2.57
6. Interiors	Walls and ceilings — 1/2″ and 5/8″ taped and finished gypsum wallboard, primed and painted with one coat latex. Pine interior trim with one coat paint or stain. Flooring — 65% carpet, 27% vinyl, 4% hardwood and 4% ceramic tile.	7.77	6.41	14.18
7. Specialties	Hardwood faced particle board case kitchen cabinets and bathroom vanities with plastic laminate countertops. Washer, dryer, drop in range with hood, dishwasher and refrigerator. Two masonry fireplaces.	5.67	2.74	8.41
8. Mechanical	Oil fired forced hot air heat with central air conditioning. One full bath, one 1/2 bath and one 3/4 bath. Stainless steel kitchen sink with disposal.	4.15	2.86	7.01
9. Electrical	200 amp service, branch circuit wiring with romex cable. Exterior and interior lighting fixtures, receptacles and switches.	2.21	1.20	3.41
10. Overhead	Contractor's overhead and profit.	3.23	2.08	5.31
	Total Cost per Square Foot	$49.43	$31.73	$81.16

To purchase a full set of Sepias, Bill of Materials and Detailed Costs — turn to page 267.

RM-3372

1½ Story
Cape Cod

2201 Square Feet
3 Bedrooms
2½ Baths
Schedule C

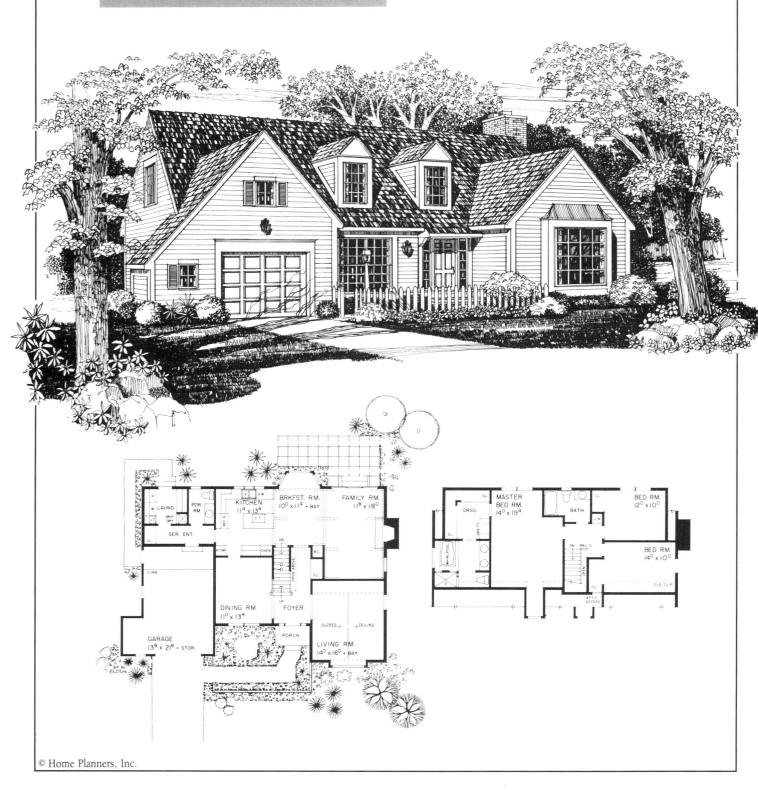

© Home Planners, Inc.

RM-3372
Cost Estimate

Cost at a Glance

Cost per Square Foot: $82.38
Total Cost: $181,318

Cost by Category

		Cost per Square Foot of Living Area		
		Materials	Installation	Total
1. Site Work	Excavation for the basement and footings.		.76	.76
2. Foundation	Main house — 10″ wide reinforced concrete foundation wall on 20″ x 10″ reinforced concrete perimeter footings. Trench footings — 10″ wide reinforced concrete walls. Slabs — 4″ thick steel trowel finished reinforced concrete over compacted gravel.	3.41	4.37	7.78
3. Framing	Exterior walls — 2 x 6 studs, 16″ on center with 1/2″ plywood sheathing. Garage — 2 x 4 studs, 16″ on center. Floor — 2 x 10 joists, 16″ on center with 3/4″ plywood subfloor. Roof — pre-engineered trusses and site cut rafters with 1/2″ plywood sheathing.	9.59	5.37	14.96
4. Exterior Walls	Beveled cedar siding over 15# felt vapor barrier with R-19 and R-11 insulation. Vinyl clad fixed, double hung and casement windows, and sliding glass patio doors.	10.15	3.99	14.14
5. Roofing	Heavyweight three tab asphalt shingles over 30# felt roofing paper. Metal roofing over bay windows. Aluminum gutters, downspouts, drip edge and flashings.	1.53	1.22	2.75
6. Interiors	Walls and ceilings — 1/2″ and 5/8″ taped and finished gypsum wallboard, primed and painted with one coat latex. Pine interior trim with one coat paint or stain. Flooring — 70% carpet, 20% vinyl, 4% hardwood and 6% ceramic tile.	7.37	6.08	13.45
7. Specialties	Hardwood faced particle board case kitchen cabinets and bathroom vanities with plastic laminate countertops. Washer, dryer, cooktop with hood, double ovens, dishwasher and refrigerator. One masonry fireplace.	9.19	2.52	11.71
8. Mechanical	Oil fired forced hot air heat with central air conditioning. One full bath, one 1/2 bath and a master suite with a whirlpool and shower. Stainless steel kitchen sink with disposal.	4.90	3.07	7.97
9. Electrical	200 amp service, branch circuit wiring with romex cable. Exterior and interior lighting fixtures, receptacles and switches.	2.20	1.27	3.47
10. Overhead	Contractor's overhead and profit.	3.38	2.01	5.39
Total Cost per Square Foot		**$51.72**	**$30.66**	**$82.38**

To purchase a full set of Sepias, Bill of Materials and Detailed Costs — turn to page 267.

RM-2563

1½ Story Cape Cod

2190 Square Feet
3 Bedrooms
2 Baths
Schedule B

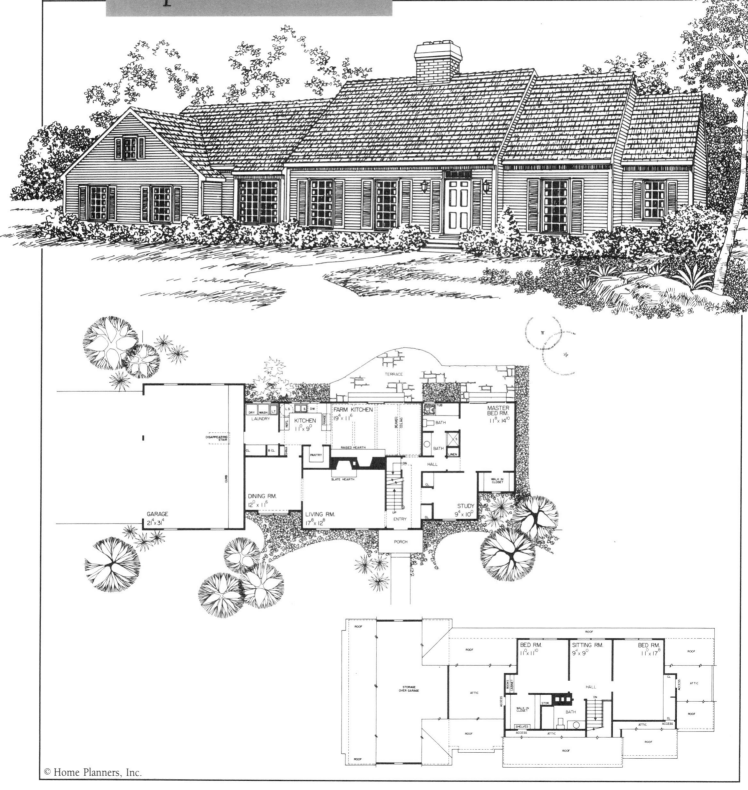

RM-2563
Cost Estimate

Cost at a Glance

Cost per Square Foot: $82.09
Total Cost: $179,777

Cost by Category

		Cost per Square Foot of Living Area		
		Materials	Installation	Total
1. Site Work	Excavation for the basement and footings.		.84	.84
2. Foundation	Main house — 12″ concrete block foundation wall on 20″ x 10″ reinforced concrete perimeter footings. Garage — 8″ x 42″ reinforced concrete trench footing. Slabs — 4″ thick reinforced steel trowel finished concrete over compacted gravel.	3.59	4.77	8.36
3. Framing	Exterior walls — 2 x 4 and 2 x 6 studs, 16″ on center with 1/2″ plywood sheathing. Floors — 2 x 10 and 2 x 12 floor joists, 16″ on center with 3/4″ plywood subfloor. Roof — pre-engineered trusses and site cut 2 x 6 rafters with 1/2″ plywood sheathing.	9.83	5.61	15.44
4. Exterior Walls	Beveled cedar siding over 15# felt vapor barrier with R-19 and R-11 wall insulation. Vinyl clad double hung and casement windows and sliding glass doors.	12.47	4.39	16.86
5. Roofing	Heavyweight three tab asphalt shingles over 30# felt roofing paper. Aluminum gutters, downspouts, drip edge and flashings.	1.57	1.35	2.92
6. Interiors	Walls and ceilings — 1/2″ and 5/8″ taped and finished gypsum wallboard, primed and painted with one coat latex. Pine interior trim. Flooring — 65% carpet, 20% vinyl, 10% ceramic tile, and 5% hardwood.	7.34	6.04	13.38
7. Specialties	Hardwood faced particle board case kitchen cabinets and bathroom vanities with plastic laminate countertops. Washer, dryer, range with hood, dishwasher, and refrigerator. Two masonry fireplaces.	5.73	3.47	9.20
8. Mechanical	Oil fired forced hot air heat with central air conditioning. Two full baths. Stainless steel double bowl kitchen sink with disposal.	3.94	2.59	6.53
9. Electrical	200 amp service, branch circuit wiring with romex cable. Exterior and interior lighting fixtures, receptacles and switches.	1.97	1.22	3.19
10. Overhead	Contractor's overhead and profit.	3.25	2.12	5.37
Total Cost per Square Foot		$49.69	$32.40	$82.09

To purchase a full set of Sepias, Bill of Materials and Detailed Costs — turn to page 267.

RM-2927
1½ Story Contemporary

2129 Square Feet
3 Bedrooms
2 Baths
Schedule B

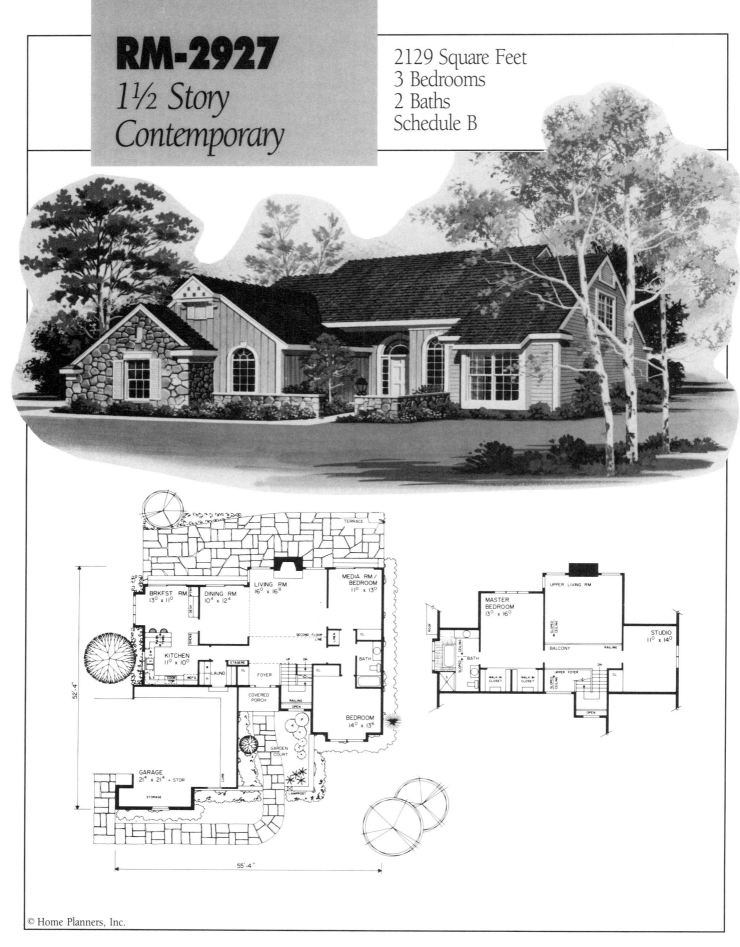

RM-2927
Cost Estimate

Cost at a Glance

Cost per Square Foot: $91.98
Total Cost: $195,825

Cost by Category

		Cost per Square Foot of Living Area		
		Materials	Installation	Total
1. Site Work	Excavation for the basement and footings.		.84	.84
2. Foundation	Main house—10″ thick reinforced concrete wall on a 20″ x 10″ reinforced concrete footing. Garage—10″ x 42″ reinforced concrete trench wall. Slabs—4″ thick reinforced steel trowel finished concrete on 4″ compacted gravel.	3.90	4.99	8.89
3. Framing	Main house—2 x 6 studs, 16″ on center with 1/2″ plywood sheathing. Garage—2 x 4 studs, 16″ on center. Floor—2 x 12 joists, 16″ on center with 3/4″ tongue and groove plywood subfloor. Roof—site cut 2 x 12, 2 x 8 and 2 x 6 rafters with 1/2″ plywood sheathing.	10.37	6.13	16.50
4. Exterior Walls	Horizontal bevel cedar siding and Texture 1-11, over 15# felt vapor barrier. Stone veneer highlights. Vinyl clad fixed, double hung and casement windows and sliding glass doors. Paneled entry door. R-19 and R-11 insulation.	16.35	7.48	23.83
5. Roofing	Heavyweight three tab asphalt roof shingles on 30# felt paper. Aluminum drip edge, flashings, gutters and downspouts.	1.75	1.47	3.22
6. Interiors	Wall finish—one coat primer and one coat paint on 1/2″ or 5/8″ gypsum wallboard. Pine door, window and baseboard moldings. Flooring—74% carpet, 15% vinyl, 5% ceramic tile and 6% hardwood.	7.18	6.42	13.60
7. Specialties	Hardwood faced, particle board case kitchen cabinets and bath vanities with plastic laminate countertops. Washer, dryer, cooktop with hood, double ovens, dishwasher and refrigerator. A masonry fireplace and terrace.	5.87	2.35	8.22
8. Mechanical	Oil fired forced hot air heat with central air conditioning. One full bath and a master suite including a whirlpool tub and shower. Double bowl kitchen sink with disposal.	4.40	2.64	7.04
9. Electrical	200 amp service, branch circuit wiring with romex cable. Exterior and interior lighting fixtures, receptacles and switches.	2.45	1.37	3.82
10. Overhead	Contractor's overhead and profit.	3.66	2.36	6.02
	Total Cost per Square Foot	$55.93	$36.05	$91.98

To purchase a full set of Sepias, Bill of Materials and Detailed Costs—turn to page 267.

RM-3476
1½ Story Contemporary

2008 Square Feet
3 Bedrooms
2½ Baths
Schedule B

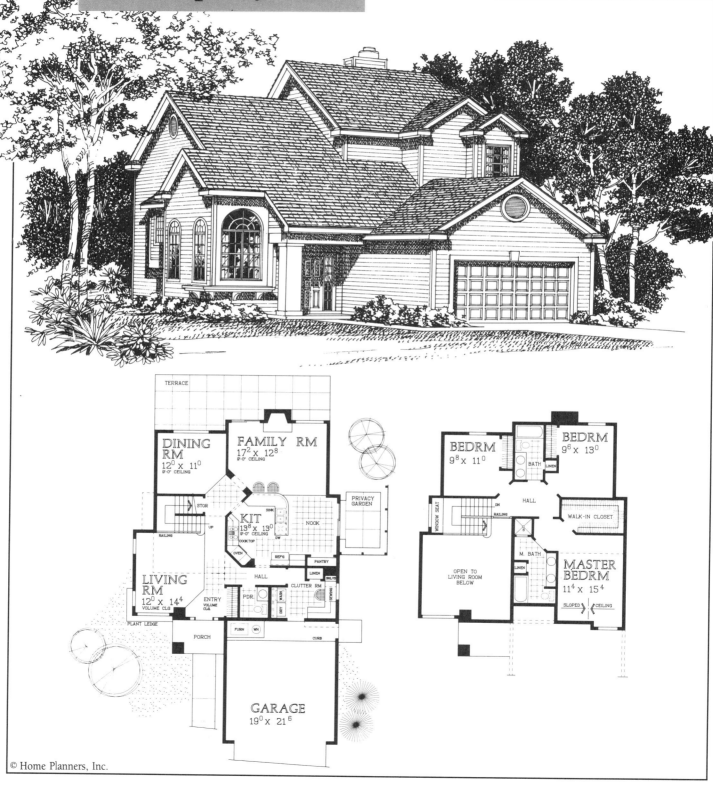

RM-3476
Cost Estimate

Cost at a Glance
Cost per Square Foot: $87.68
Total Cost: $176,061

Cost by Category

		Cost per Square Foot of Living Area		
		Materials	Installation	Total
1. Site Work	Excavation for the slab and footings.		.32	.32
2. Foundation	Main house — 6" wide reinforced concrete foundation wall on 16" x 10" reinforced concrete perimeter footings. Slabs — 4" thick steel trowel finished reinforced concrete over compacted gravel.	2.29	2.89	5.18
3. Framing	Exterior walls — 2 x 6 studs, 16" on center with 1/2" plywood sheathing. Floor — 2 x 10 floor joists, 16" on center with 3/4" plywood subfloor. Roof — pre-engineered trusses and site cut rafters with 5/8" plywood sheathing.	12.29	12.25	24.54
4. Exterior Walls	Beveled cedar siding over 15# felt vapor barrier with R-19 and R-11 insulation. Vinyl clad fixed, double hung and sliding windows and sliding glass patio doors.	10.57	4.15	14.72
5. Roofing	Heavyweight three tab asphalt shingles over 30# felt roofing paper. Aluminum gutters, downspouts, drip edge and flashings.	1.63	1.41	3.04
6. Interiors	Walls and ceilings — 1/2" and 5/8" taped and finished gypsum wallboard, primed and painted with one coat latex. Pine interior trim with one coat paint or stain. Flooring — 58% carpet, 6% vinyl, and 36% ceramic tile.	7.28	6.56	13.84
7. Specialties	Hardwood faced particle board case kitchen cabinets and bathroom vanities with plastic laminate countertops. Washer, dryer, cooktop with hood, double ovens, dishwasher and refrigerator. One pre-manufactured fireplace.	6.95	1.78	8.73
8. Mechanical	Oil fired forced hot air heat with central air conditioning. One full bath, one 1/2 bath and a master suite with a tub and shower. Stainless steel kitchen sink with disposal.	4.46	3.02	7.48
9. Electrical	200 amp service, branch circuit wiring with romex cable. Exterior and interior lighting fixtures, receptacles and switches.	2.63	1.46	4.09
10. Overhead	Contractor's overhead and profit.	3.37	2.37	5.74
	Total Cost per Square Foot	**$51.47**	**$36.21**	**$87.68**

To purchase a full set of Sepias, Bill of Materials and Detailed Costs — turn to page 267.

RM-3455

1½ Story Contemporary

2075 Square Feet
3 Bedrooms
2½ Baths
Schedule B

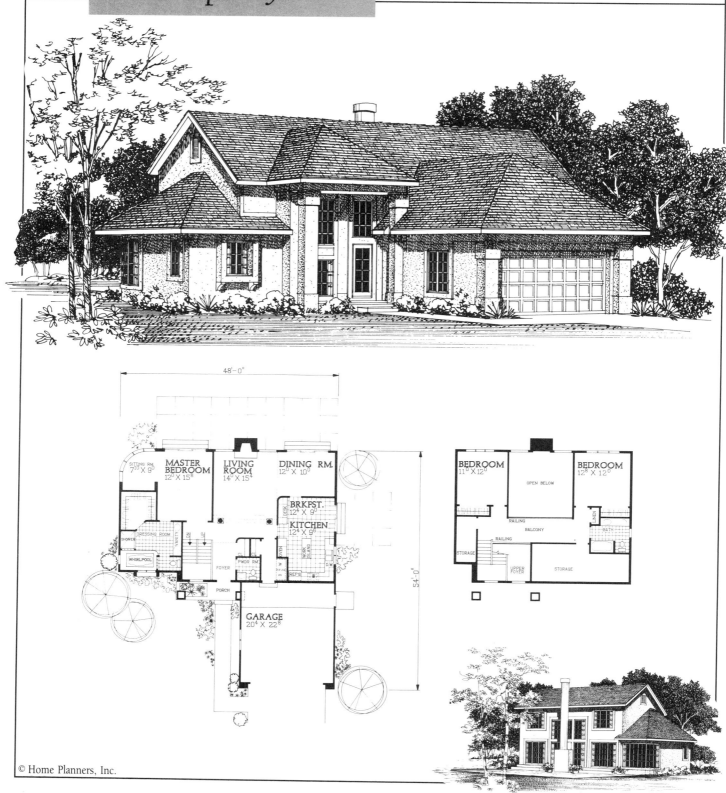

48'-0"

54'-0"

SITTING RM.
7¹⁰ X 9⁰

MASTER BEDROOM
12⁰ X 15⁸

LIVING ROOM
14⁴ X 15⁴

DINING RM.
12⁰ X 10⁰

DRESSING ROOM

SHOWER

WHIRLPOOL

VANITY

FOYER

PWDR RM

CL. CL.

BRKFST.
12⁴ X 9⁰

KITCHEN
12⁴ X 9⁰

WORK ISLAND

OVEN

REFG.

D.W.

PORCH

GARAGE
20⁴ X 22⁸

BEDROOM
11⁰ X 12⁰

BEDROOM
12⁸ X 12⁰

OPEN BELOW

RAILING

BALCONY

RAILING

LINEN

BATH

STORAGE

UPPER FOYER

STORAGE

RM-3455
Cost Estimate

Cost by Category

		Cost per Square Foot of Living Area		
		Materials	Installation	Total
1. Site Work	Excavation for the basement and footings.		.85	.85
2. Foundation	Main house — 10″ wide reinforced concrete foundation wall on 20″ x 10″ reinforced concrete perimeter footings. Trench footings — 8″, 10″ and 14″ wide reinforced concrete. Slabs — 4″ thick steel trowel finished reinforced concrete over compacted gravel.	3.81	4.63	8.44
3. Framing	Exterior walls — 2 x 6 studs, 16″ on center with 1/2″ plywood sheathing. Garage — 2 x 4 studs, 16″ on center. Floor — 2 x 12 floor joists, 16″ on center with 3/4″ plywood subfloor. Roof — pre-engineered trusses and site cut rafters with 5/8″ plywood sheathing.	10.86	6.83	17.69
4. Exterior Walls	1″ thick stucco siding on 3/8″ high rib metal lath over 15# felt vapor barrier with R-19 and R-11 insulation. Vinyl clad fixed and casement windows and sliding glass patio doors.	10.41	2.31	12.72
5. Roofing	Heavyweight three tab asphalt shingles over 30# felt roofing paper. Aluminum gutters, downspouts, drip edge and flashings.	1.81	1.51	3.32
6. Interiors	Walls and ceilings — 1/2″ and 5/8″ taped and finished gypsum wallboard, primed and painted with one coat latex. Pine interior trim with one coat paint or stain. Flooring — 68% carpet, 16% vinyl, 6% hardwood and 10% ceramic tile.	6.79	6.13	12.92
7. Specialties	Hardwood faced particle board case kitchen cabinets and bathroom vanities with plastic laminate countertops. Washer, dryer, cooktop with hood, double ovens, dishwasher and refrigerator. One pre-manufactured fireplace.	6.76	1.98	8.74
8. Mechanical	Oil fired forced hot air heat with central air conditioning. One full bath, one 1/2 bath and a master suite with a whirlpool tub and shower. Stainless steel kitchen sink with disposal.	4.97	3.11	8.08
9. Electrical	200 amp service, branch circuit wiring with romex cable. Exterior and interior lighting fixtures, receptacles and switches.	2.73	1.46	4.19
10. Overhead	Contractor's overhead and profit.	3.37	2.02	5.39
Total Cost per Square Foot		$51.51	$30.83	$82.34

To purchase a full set of Sepias, Bill of Materials and Detailed Costs — turn to page 267.

RM-3338
1½ Story Contemporary

2284 Square Feet
3 Bedrooms
2½ Baths
Schedule B

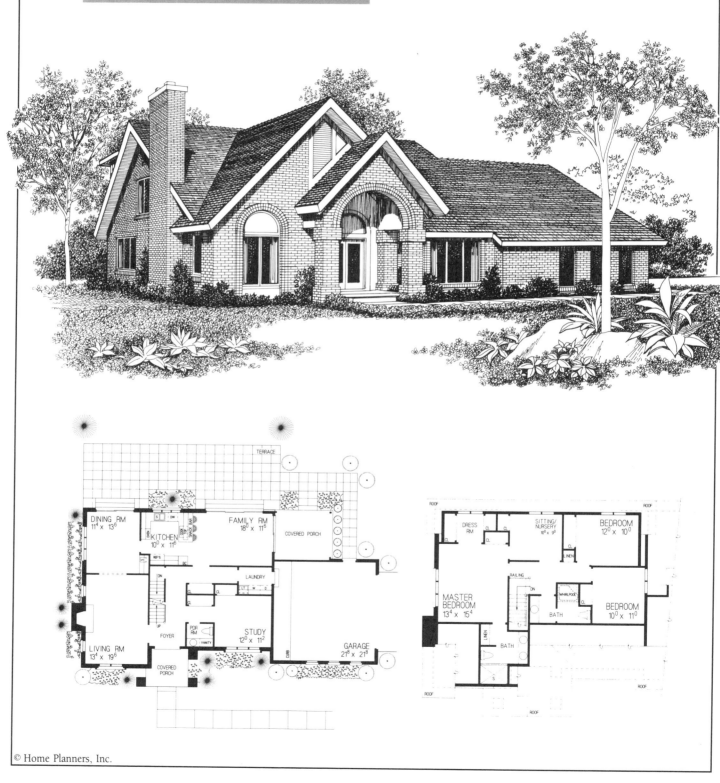

RM-3338
Cost Estimate

Cost at a Glance

Cost per Square Foot: $80.82
Total Cost: $184,592

Cost by Category

		Cost per Square Foot of Living Area		
		Materials	Installation	Total
1. Site Work	Excavation for the basement and footings.		.75	.75
2. Foundation	Main house — 10″ wide reinforced concrete foundation wall on 20″ x 10″ reinforced concrete perimeter footings. Trench footings — 8″ and 10″ wide reinforced concrete. Slabs — 4″ thick steel trowel finished reinforced concrete over compacted gravel.	3.67	4.59	8.26
3. Framing	Exterior walls — 2 x 4 studs, 16″ on center with 1/2″ plywood sheathing. Garage — 2 x 4 studs, 16″ on center. Floor — 2 x 10 and 2 x 12 joists, 16″ on center with 3/4″ plywood subfloor. Roof — site cut rafters with 1/2″ plywood sheathing.	9.75	5.76	15.51
4. Exterior Walls	Brick veneer and Texture 1-11 siding over 15# felt vapor barrier with R-19 and R-11 insulation. Vinyl clad fixed and casement windows and sliding glass patio doors.	10.84	6.03	16.87
5. Roofing	Heavyweight three tab asphalt shingles over 30# felt roofing paper. Aluminum gutters, downspouts, drip edge and flashings.	1.39	1.13	2.52
6. Interiors	Walls and ceilings — 1/2″ and 5/8″ taped and finished gypsum wallboard, primed and painted with one coat latex. Pine interior trim. Flooring — 77% carpet, 16% vinyl, 4% hardwood and 3% ceramic tile.	7.21	6.40	13.61
7. Specialties	Hardwood faced particle board case kitchen cabinets and bathroom vanities with plastic laminate countertops. Washer, dryer, cooktop, double wall ovens, dishwasher and refrigerator. One masonry fireplace.	4.85	2.10	6.95
8. Mechanical	Oil fired forced hot air heat with central air conditioning. One full bath, one 1/2 bath and one 3/4 bath. Stainless steel kitchen sink with disposal.	4.52	2.84	7.36
9. Electrical	200 amp service, branch circuit wiring with romex cable. Exterior and interior lighting fixtures, receptacles and switches.	2.37	1.33	3.70
10. Overhead	Contractor's overhead and profit.	3.12	2.17	5.29
	Total Cost per Square Foot	$47.72	$33.10	$80.82

To purchase a full set of Sepias, Bill of Materials and Detailed Costs — turn to page 267.

RM-3458
1½ Story Contemporary

2343 Square Feet
4 Bedrooms
2½ Baths
Schedule C

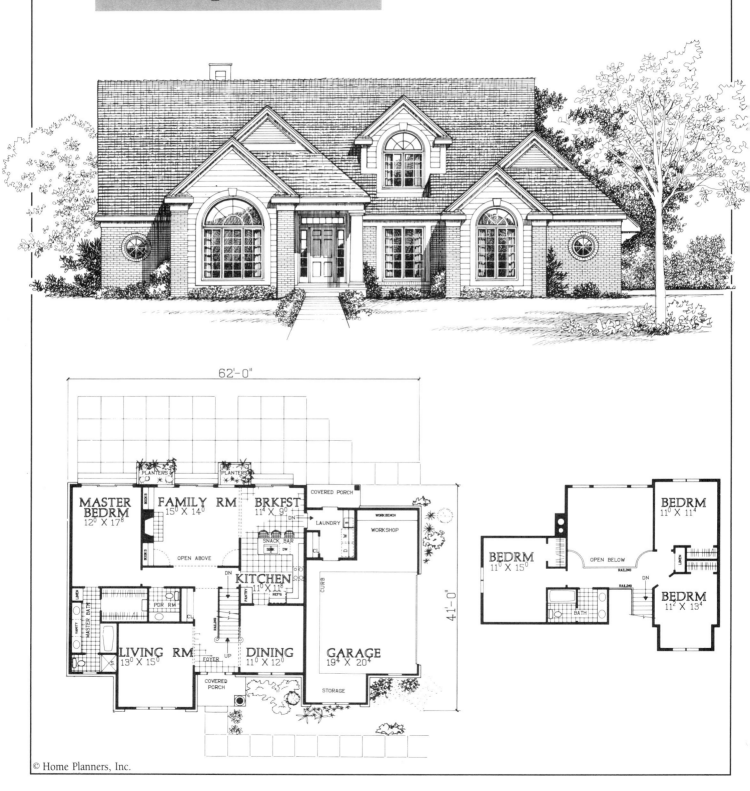

RM-3458
Cost Estimate

Cost at a Glance

Cost per Square Foot: $90.30
Total Cost: $211,572

Cost by Category

		Cost per Square Foot of Living Area		
		Materials	Installation	Total
1. Site Work	Excavation for the basement and footings.		.81	.81
2. Foundation	Main house — 10″ wide reinforced concrete foundation wall on 18″ x 10″ reinforced concrete perimeter footings. Trench footings — 8″, 10″ and 14″ wide reinforced concrete. Slabs — 4″ thick steel trowel finished reinforced concrete over compacted gravel.	3.67	4.61	8.28
3. Framing	Exterior walls — 2 x 6 studs, 16″ on center with 1/2″ plywood sheathing. Garage — 2 x 4 studs, 16″ on center. Floor — 2 x 12 floor joists, 16″ on center with 3/4″ plywood subfloor. Roof — pre-engineered trusses and site cut rafters with 1/2″ plywood sheathing.	10.60	6.23	16.83
4. Exterior Walls	Beveled cedar and brick veneer siding over 15# felt vapor barrier with R-19 and R-11 insulation. Vinyl clad fixed and casement windows and sliding glass patio doors.	13.57	8.26	21.83
5. Roofing	Heavyweight three tab asphalt shingles over 30# felt roofing paper. Aluminum gutters, downspouts, drip edge and flashings.	1.95	1.59	3.54
6. Interiors	Walls and ceilings — 1/2″ and 5/8″ taped and finished gypsum wallboard, primed and painted with one coat latex. Pine interior trim with one coat paint or stain. Flooring — 68% carpet, 15% vinyl, 7% hardwood and 10% ceramic tile.	7.09	6.27	13.36
7. Specialties	Hardwood faced particle board case kitchen cabinets and bathroom vanities with plastic laminate countertops. Washer, dryer, cooktop with hood, double ovens, dishwasher and refrigerator. One pre-manufactured fireplace.	6.65	1.85	8.50
8. Mechanical	Oil fired forced hot air heat with central air conditioning. One full bath, one 1/2 bath and a master suite with a tub and shower. Stainless steel kitchen sink with disposal.	4.75	3.02	7.77
9. Electrical	200 amp service, branch circuit wiring with romex cable. Exterior and interior lighting fixtures, receptacles and switches.	2.20	1.28	3.48
10. Overhead	Contractor's overhead and profit.	3.53	2.37	5.90
	Total Cost per Square Foot	$54.01	$36.29	$90.30

To purchase a full set of Sepias, Bill of Materials and Detailed Costs — turn to page 267.

RM-3461
1½ Story Farmhouse

2002 Square Feet
4 Bedrooms
2½ Baths
Schedule B

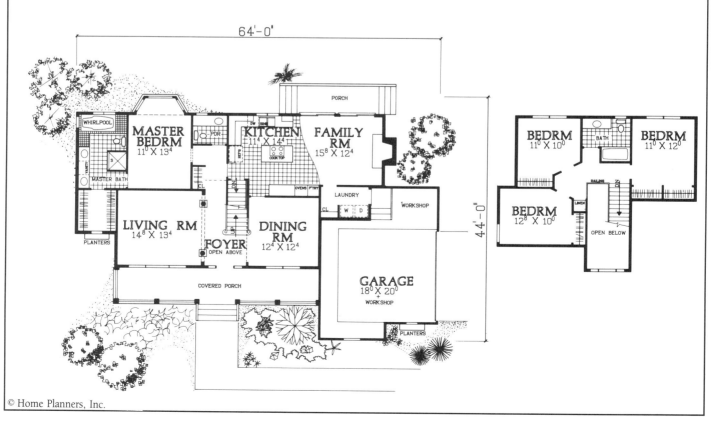

64'-0"

44'-0"

WHIRLPOOL

MASTER BEDRM
11⁰ X 13⁴

VANITY

MASTER BATH

PLANTERS

LIVING RM
14⁸ X 13⁴

PDR

FOYER
OPEN ABOVE

KITCHEN
11⁴ X 14⁴

COOK TOP

DW SINK

REF.

OVENS PTRY

DINING RM
12⁴ X 12⁴

CL

FAMILY RM
15⁸ X 12⁴

PORCH

LAUNDRY

CL   W   D

WORKSHOP

GARAGE
18⁰ X 20⁰
+
WORKSHOP

PLANTERS

COVERED PORCH

BEDRM
11⁰ X 10⁰

BATH

BEDRM
11⁰ X 12⁰

BEDRM
12⁸ X 10⁰

LINEN

RAILING

DN

OPEN BELOW

RM-3461
Cost Estimate

Cost at a Glance

Cost per Square Foot: $94.40
Total Cost: $188,988

Cost by Category

		Cost per Square Foot of Living Area		
		Materials	Installation	Total
1. Site Work	Excavation for the basement and footings.		.88	.88
2. Foundation	Main house — 10″ wide reinforced concrete foundation wall on 20″ x 10″ reinforced concrete perimeter footings. Garage — 8″ x 42″ reinforced concrete trench footing. Slabs — 4″ thick reinforced steel trowel finished concrete over compacted gravel.	3.95	5.01	8.96
3. Framing	Exterior walls — 2 x 6 studs, 16″ on center with 1/2″ plywood sheathing. Floors — 2 x 10 and 2 x 12 floor joists, 16″ on center with 3/4″ plywood subfloor. Garage — 2 x 4 studs, 16″ on center with 1/2″ plywood sheathing. Roof — pre-engineered trusses and site cut 2 x 6 rafters with 1/2″ plywood sheathing.	11.33	7.01	18.34
4. Exterior Walls	Beveled cedar siding over 15# felt vapor barrier with R-19 and R-11 wall insulation. Vinyl fixed and casement windows and sliding glass doors. Paneled front door.	15.02	5.01	20.03
5. Roofing	Heavyweight three tab asphalt shingles over 30# felt roofing paper. Aluminum gutters, downspouts, drip edge and flashings.	2.51	1.71	4.22
6. Interiors	Walls and ceilings — 1/2″ and 5/8″ taped and finished gypsum wallboard, primed and painted with one coat latex. Pine interior trim with one coat paint or stain. Flooring — 77% carpet, 15% vinyl, 2% ceramic tile, and 6% hardwood.	7.99	6.78	14.77
7. Specialties	Hardwood faced particle board case kitchen cabinets and bathroom vanities with plastic laminate countertops. Washer, dryer, cooktop with hood, double ovens, dishwasher, and refrigerator. One prefabricated fireplace.	6.99	1.94	8.93
8. Mechanical	Oil fired forced hot air heat with central air conditioning. One full bath, one 1/2 bath and a master suite with a whirlpool tub and shower. Stainless steel double bowl kitchen sink with disposal.	5.11	3.16	8.27
9. Electrical	200 amp service, branch circuit wiring with romex cable. Exterior and interior lighting fixtures, receptacles and switches.	2.41	1.42	3.83
10. Overhead	Contractor's overhead and profit.	3.87	2.30	6.17
Total Cost per Square Foot		$59.18	$35.22	$94.40

To purchase a full set of Sepias, Bill of Materials and Detailed Costs — turn to page 267.

RM-3318A

1½ Story Traditional

2097 Square Feet
2 Bedrooms
2 Baths
Schedule B

© Home Planners, Inc.

RM-3318A
Cost Estimate

Cost at a Glance

Cost per Square Foot: $90.36
Total Cost: $189,484

Cost by Category

		Cost per Square Foot of Living Area		
		Materials	Installation	Total
1. Site Work	Excavation for the basement and footings.		.68	.68
2. Foundation	Main house — 10″ wide reinforced concrete foundation wall on 20″ x 10″ reinforced concrete perimeter footings. Trench footings — 8″ and 16″ wide reinforced concrete. Slabs — 4″ thick steel trowel finished reinforced concrete over compacted gravel.	4.91	5.84	10.75
3. Framing	Exterior walls — 2 x 6 studs, 16″ on center with 1/2″ plywood sheathing. Floor — 2 x 12 joists, 16″ on center with 3/4″ plywood subfloor. Roof — pre-engineered trusses and site cut rafters with 5/8″ plywood sheathing.	12.25	6.34	18.59
4. Exterior Walls	Beveled cedar and fieldstone veneer siding over 15# felt vapor barrier with R-19 insulation. Vinyl clad fixed and casement windows with swinging and sliding glass patio doors.	15.01	5.59	20.60
5. Roofing	Heavyweight three tab asphalt shingles over 30# felt roofing paper. Aluminum gutters, downspouts, drip edge and flashings.	1.38	1.17	2.55
6. Interiors	Walls and ceilings — 1/2″ and 5/8″ taped and finished gypsum wallboard, primed and painted with one coat latex. Pine interior trim, with one coat paint or stain. Flooring — 69% carpet, 18% vinyl, 8% hardwood and 5% ceramic tile.	6.34	5.79	12.13
7. Specialties	Hardwood faced particle board case kitchen cabinets and bathroom vanities with plastic laminate countertops. Washer, dryer, cooktop with hood, double wall ovens, dishwasher and refrigerator. Two double faced pre-fabricated fireplaces.	6.28	1.66	7.94
8. Mechanical	Oil fired forced hot air heat with central air conditioning. One full bath, and a master suite with a whirlpool and shower. Stainless steel kitchen sink with disposal.	4.88	3.01	7.89
9. Electrical	200 amp service, branch circuit wiring with romex cable. Exterior and interior lighting fixtures, receptacles and switches.	2.14	1.18	3.32
10. Overhead	Contractor's overhead and profit.	3.72	2.19	5.91
Total Cost per Square Foot		$56.91	$33.45	$90.36

To purchase a full set of Sepias, Bill of Materials and Detailed Costs — *turn to page 267.*

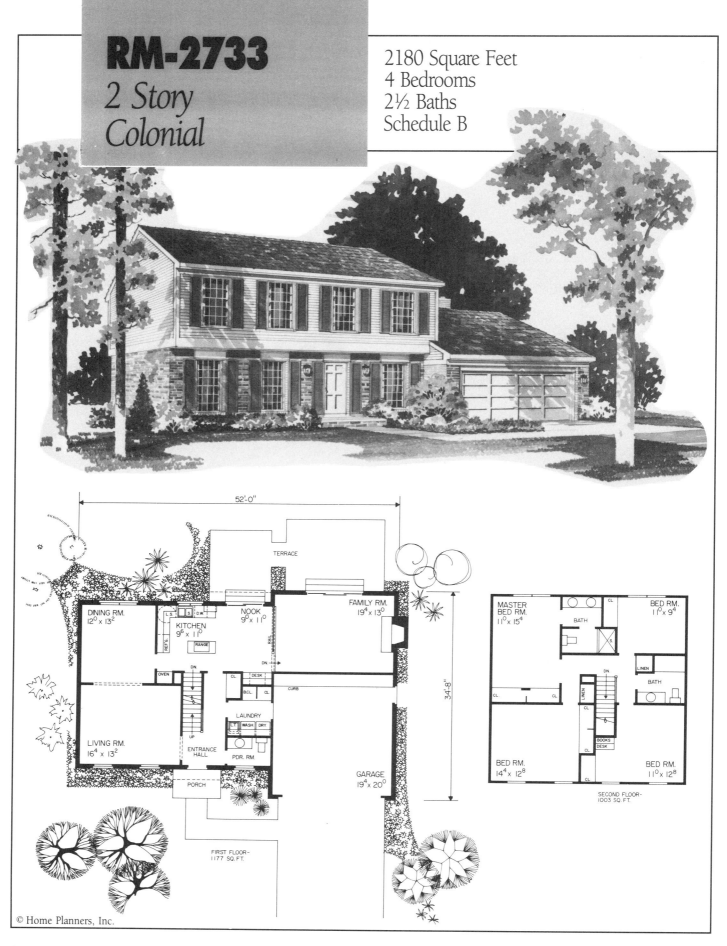

RM-2733

2 Story Colonial

2180 Square Feet
4 Bedrooms
2½ Baths
Schedule B

52'-0"

34'-8"

TERRACE

DINING RM.
12⁰ x 13²

KITCHEN
9⁶ x 11⁰

NOOK
9⁰ x 11⁰

FAMILY RM.
19⁴ x 13⁰

L.S.   S   D.W.

REF'G.

RANGE

OVEN

DN

DESK

B.CL.   CL

CURB

DN

RAIL

CL

LIVING RM.
16⁴ x 13²

UP

ENTRANCE
HALL

LAUNDRY

WASH   DRY

PDR. RM.

PORCH

GARAGE
19⁴ x 20⁰

FIRST FLOOR—
1177 SQ. FT.

MASTER
BED RM.
11⁰ x 15⁴

BATH

S

BED RM.
11⁰ x 9⁴

LINEN

BATH

CL   CL

DN

LINEN

CL

CL

BOOKS
DESK

CL

BED RM.
14⁴ x 12⁸

BED RM.
11⁰ x 12⁸

CL

SECOND FLOOR—
1003 SQ. FT.

RM-2733
Cost Estimate

Cost at a Glance
Cost per Square Foot: $68.62
Total Cost: $149,591

Cost by Category

		Cost per Square Foot of Living Area		
		Materials	Installation	Total
1. Site Work	Excavation for the basement and footings.		.74	.74
2. Foundation	Main house — 12″ wide concrete masonry unit foundation wall on 20″ x 10″ reinforced concrete perimeter footings. Trench footings — 8″ and 12″ wide reinforced concrete. Slabs — 4″ thick steel trowel finished reinforced concrete over compacted gravel.	2.90	3.81	6.71
3. Framing	Exterior walls — 2 x 4 studs, 16″ on center with 1/2″ plywood sheathing. Floor — 2 x 10 and 2 x 12 joists, 16″ on center with 3/4″ plywood subfloor. Roof — pre-engineered trusses with 1/2″ plywood sheathing.	6.73	4.20	10.93
4. Exterior Walls	Brick veneer and beveled cedar siding over 15# felt vapor barrier with R-19 and R-11 insulation. Vinyl clad casement and double hung windows and sliding glass patio doors.	8.48	4.47	12.95
5. Roofing	Heavyweight three tab asphalt shingles over 30# felt roofing paper. Aluminum gutters, downspouts, drip edge and flashings.	1.02	.93	1.95
6. Interiors	Walls and ceilings — 1/2″ and 5/8″ taped and finished gypsum wallboard, primed and painted with one coat latex. Pine interior trim with one coat paint or stain. Flooring — 77% carpet, 16% vinyl, 4% hardwood and 3% ceramic tile.	6.79	5.78	12.57
7. Specialties	Hardwood faced particle board case kitchen cabinets and bathroom vanities with plastic laminate countertops. Washer, dryer, cooktop with hood, double wall ovens, dishwasher and refrigerator. One masonry fireplace.	5.93	2.06	7.99
8. Mechanical	Oil fired forced hot air heat with central air conditioning. One full bath, one 1/2 bath and one 3/4 bath. Stainless steel kitchen sink with disposal.	4.05	2.78	6.83
9. Electrical	200 amp service, branch circuit wiring with romex cable. Exterior and interior lighting fixtures, receptacles and switches.	2.21	1.25	3.46
10. Overhead	Contractor's overhead and profit.	2.67	1.82	4.49
	Total Cost per Square Foot	**$40.78**	**$27.84**	**$68.62**

To purchase a full set of Sepias, Bill of Materials and Detailed Costs — turn to page 267.

RM-3484

2 *Story*
Contemporary

2087 Square Feet
3 Bedrooms
2½ Baths
Schedule B

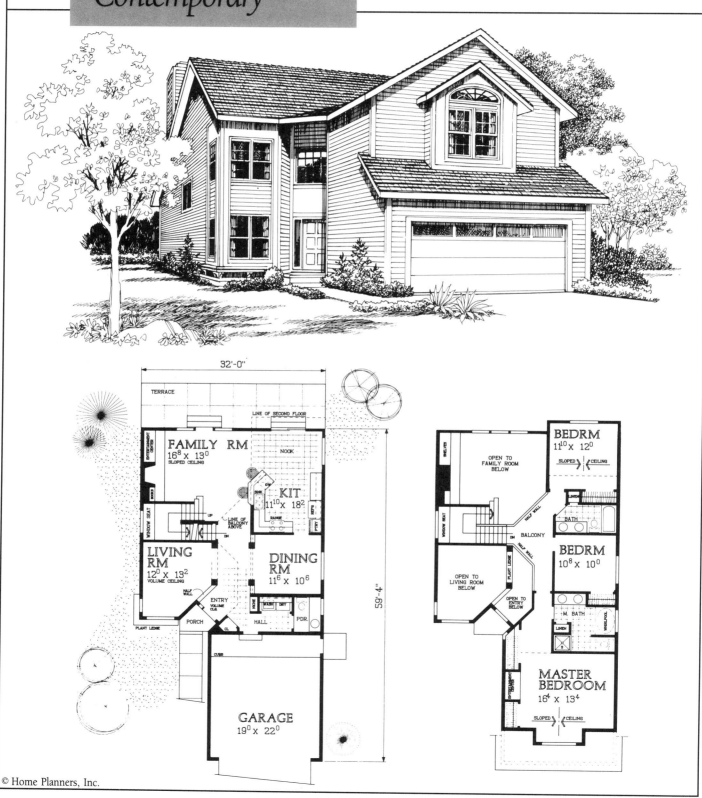

FAMILY RM
16⁸ x 13⁰
SLOPED CEILING

NOOK

KIT
11¹⁰ x 18²

LIVING RM
12⁰ x 13²
VOLUME CEILING

DINING RM
11⁶ x 10⁶

TERRACE

LINE OF SECOND FLOOR

32'-0"

59'-4"

ENTRY VOLUME CLG.

PORCH

HALL

PDR.

GARAGE
19⁰ x 22⁰

BEDRM
11¹⁰ x 12⁰
SLOPED ◁▷ CEILING

OPEN TO FAMILY ROOM BELOW

BATH

BALCONY

BEDRM
10⁸ x 10⁰

OPEN TO LIVING ROOM BELOW

OPEN TO ENTRY BELOW

M. BATH

MASTER BEDROOM
16⁴ x 13⁴
SLOPED ◁▷ CEILING

RM-3484
Cost Estimate

Cost at a Glance

Cost per Square Foot: $87.18
Total Cost: $181,944

Cost by Category

Category	Description	Cost per Square Foot of Living Area		
		Materials	Installation	Total
1. Site Work	Excavation for the basement and footings.		.76	.76
2. Foundation	Main house—8″ and 10″ wide reinforced concrete foundation walls on 20″ x 10″ reinforced concrete perimeter footings. Slabs—4″ thick steel trowel finished reinforced concrete over compacted gravel.	3.32	4.11	7.43
3. Framing	Exterior walls—2 x 6 studs, 16″ on center with 1/2″ plywood sheathing. Floor—2 x 10 joists, 16″ on center with 3/4″ plywood subfloor. Roof—pre-engineered trusses and site cut rafters with 5/8″ plywood sheathing.	10.82	8.48	19.30
4. Exterior Walls	Beveled cedar siding over 15# felt vapor barrier with R-19 and R-11 insulation. Vinyl clad fixed, sliding and double hung windows and a swinging glass patio door.	12.39	3.88	16.27
5. Roofing	Red cedar shingle roofing over 30# felt roofing paper. Aluminum gutters, downspouts, drip edge and flashings.	2.93	1.47	4.40
6. Interiors	Walls and ceilings—1/2″ and 5/8″ taped and finished gypsum wallboard, primed and painted with one coat latex. Pine interior trim with one coat paint or stain. Flooring—71% carpet, 13% vinyl, and 16% ceramic tile.	6.79	6.37	13.16
7. Specialties	Hardwood faced particle board case kitchen cabinets and bathroom vanities with plastic laminate countertops. Washer, dryer, range with hood, dishwasher and refrigerator. One pre-manufactured fireplace.	5.96	1.59	7.55
8. Mechanical	Oil fired forced hot air heat with central air conditioning. One full bath, one 1/2 bath and a master suite with a whirlpool and shower. Stainless steel kitchen sink with disposal.	4.98	3.10	8.08
9. Electrical	200 amp service, branch circuit wiring with romex cable. Exterior and interior lighting fixtures, receptacles and switches.	2.93	1.60	4.53
10. Overhead	Contractor's overhead and profit.	3.51	2.19	5.70
Total Cost per Square Foot		$53.63	$33.55	$87.18

To purchase a full set of Sepias, Bill of Materials and Detailed Costs—turn to page 267.

RM-3477
2 Story Contemporary

2152 Square Feet
3 Bedrooms
2½ Baths
Schedule C

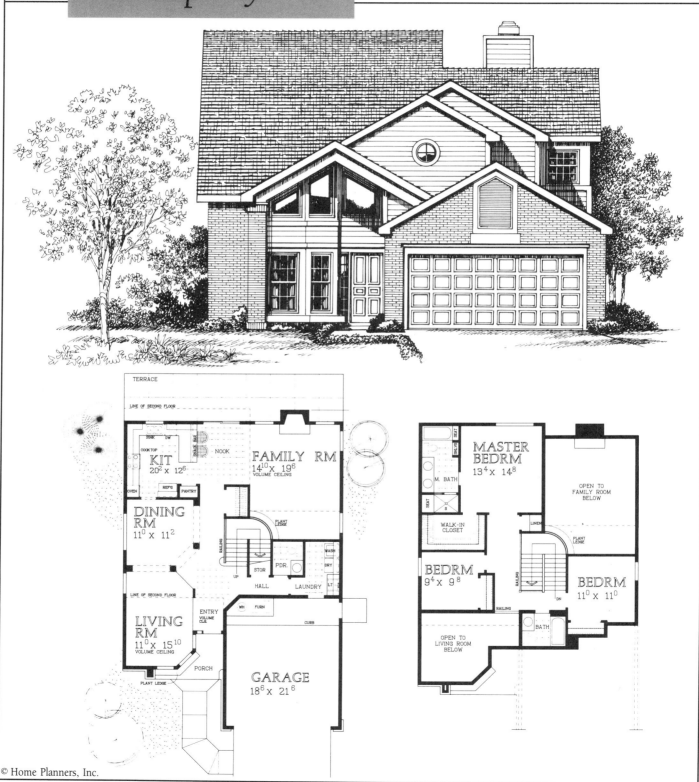

RM-3477
Cost Estimate

Cost at a Glance

Cost per Square Foot: $75.74
Total Cost: $162,992

Cost by Category

		Cost per Square Foot of Living Area		
		Materials	Installation	Total
1. Site Work	Excavation for the slab and footings.		.30	.30
2. Foundation	Main house — 6″ and 10″ wide reinforced concrete foundation walls on 16″ x 10″ and 20″ x 10″ reinforced concrete perimeter footings. Slabs — 4″ thick steel trowel finished reinforced concrete over compacted gravel.	2.12	2.58	4.70
3. Framing	Exterior walls — 2 x 6 studs, 16″ on center with 1/2″ plywood sheathing. Floor — 2 x 10 floor joists, 16″ on center with 3/4″ plywood subfloor. Roof — pre-engineered trusses and site cut rafters with 5/8″ plywood sheathing.	10.26	9.68	19.94
4. Exterior Walls	Beveled cedar and brick veneer siding over 15# felt vapor barrier with R-19 and R-11 insulation. Vinyl clad fixed, double hung and sliding windows and a sliding glass patio door.	8.91	3.38	12.29
5. Roofing	Heavyweight three tab asphalt shingles over 30# felt roofing paper. Aluminum gutters, downspouts, drip edge and flashings.	1.39	1.07	2.46
6. Interiors	Walls and ceilings — 1/2″ and 5/8″ taped and finished gypsum wallboard, primed and painted with one coat latex. Pine interior trim with one coat paint or stain. Flooring — 70% carpet, 13% vinyl, and 17% ceramic tile.	6.44	6.12	12.56
7. Specialties	Hardwood faced particle board case kitchen cabinets and bathroom vanities with plastic laminate countertops. Washer, dryer, cooktop with hood, double ovens, dishwasher and refrigerator. One pre-manufactured fireplace.	6.22	1.58	7.80
8. Mechanical	Oil fired forced hot air heat with central air conditioning. One full bath, one 1/2 bath and a master suite with a tub and shower. Stainless steel kitchen sink with disposal.	4.28	2.89	7.17
9. Electrical	200 amp service, branch circuit wiring with romex cable. Exterior and interior lighting fixtures, receptacles and switches.	2.30	1.26	3.56
10. Overhead	Contractor's overhead and profit.	2.94	2.02	4.96
Total Cost per Square Foot		$44.86	$30.88	$75.74

To purchase a full set of Sepias, Bill of Materials and Detailed Costs — turn to page 267.

RM-2776

2 Story Farmhouse

2008 Square Feet
3 Bedrooms
2½ Baths
Schedule B

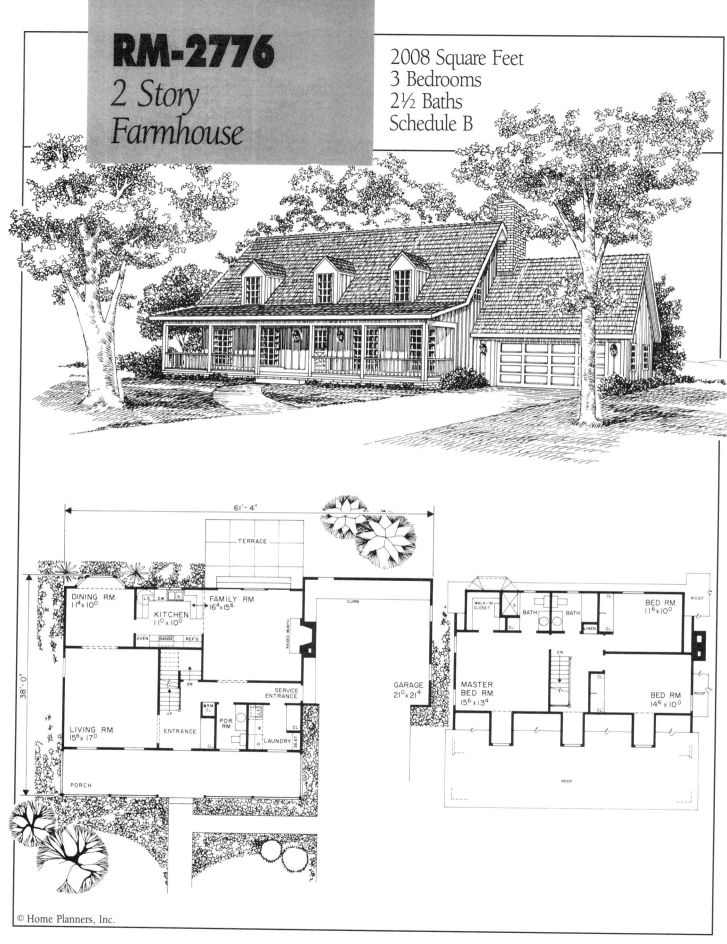

61'- 4"

TERRACE

DINING RM.
11⁴ x 10⁰

KITCHEN
11⁰ x 10⁰

L.S. DW. S

FAMILY RM.
16⁴ x 15⁶

CURB

RAISED HEARTH

OVEN   RANGE   REF'G

38'- 0"

DN

UP

LIVING RM.
15⁶ x 17⁰

RM CL

PDR. RM

ENTRANCE

CL

SERVICE ENTRANCE

LAUNDRY

SEAT

GARAGE
21⁰ x 21⁴

PORCH

WALK-IN CLOSET

S

CL

BATH   BATH

LINEN   CL

CL

BED RM.
11⁶ x 10⁰

ROOF

DN

MASTER BED RM.
15⁶ x 13⁴

CL

CL

BED RM.
14⁶ x 10⁰

ROOF

ROOF

RM-2776
Cost Estimate

Cost at a Glance

Cost per Square Foot: *$77.67*
Total Cost: *$155,961*

Cost by Category

		Cost per Square Foot of Living Area		
		Materials	Installation	Total
1. Site Work	Excavation for the basement and footings.		.79	.79
2. Foundation	Main house — 12″ concrete block wall on a 20″ x 10″ reinforced concrete footing. Garage — 8″ x 42″ reinforced concrete foundation wall. Slabs — 4″ thick reinforced steel trowel finished concrete on 4″ compacted gravel.	3.07	3.82	6.89
3. Framing	Main house — 2 x 4 studs, 16″ on center with 1/2″ plywood sheathing. Garage — 2 x 4 studs, 16″ on center. Floor — 2 x 12 joists, 16″ on center with 3/4″ tongue and groove plywood subfloor. Roof — site cut 2 x 6 and 2 x 8 rafters with 1/2″ plywood sheathing.	10.22	5.84	16.06
4. Exterior Walls	Exterior plywood/batten siding with 15# felt vapor barrier. Vinyl clad double hung, fixed and casement windows and sliding glass doors, crossbuck entry doors. R-19 and R-11 insulation.	8.57	4.28	12.85
5. Roofing	Heavyweight three tab asphalt roof shingles on 30# felt paper. Aluminum drip edge, flashings, gutters and downspouts.	1.94	1.54	3.48
6. Interiors	Wall finish — one coat primer and one coat paint on 1/2″ or 5/8″ gypsum wallboard. Pine door, window and baseboard moldings, painted or stained one coat. Flooring — 80% carpet, 9% vinyl, 5% ceramic tile and 6% hardwood.	7.60	6.46	14.06
7. Specialties	Hardwood faced, particle board case kitchen cabinets and bath vanities with plastic laminate countertops. Washer, dryer, cooktop with hood, double ovens, dishwasher and refrigerator. One masonry fireplace, covered porch and terrace.	5.77	2.56	8.33
8. Mechanical	Oil fired forced hot air heat with central air conditioning. One full bath, one 3/4 bath and one half bath. Double bowl kitchen sink with disposal.	4.11	2.78	6.89
9. Electrical	200 amp service, branch circuit wiring with romex cable. Exterior and interior lighting fixtures, receptacles and switches.	2.08	1.15	3.23
10. Overhead	Contractor's overhead and profit.	3.04	2.05	5.09
	Total Cost per Square Foot	**$46.40**	**$31.27**	**$77.67**

To purchase a full set of Sepias, Bill of Materials and Detailed Costs — turn to page 267.

RM-2774

2 Story Farmhouse

2339 Square Feet
4 Bedrooms
2½ Baths
Schedule B

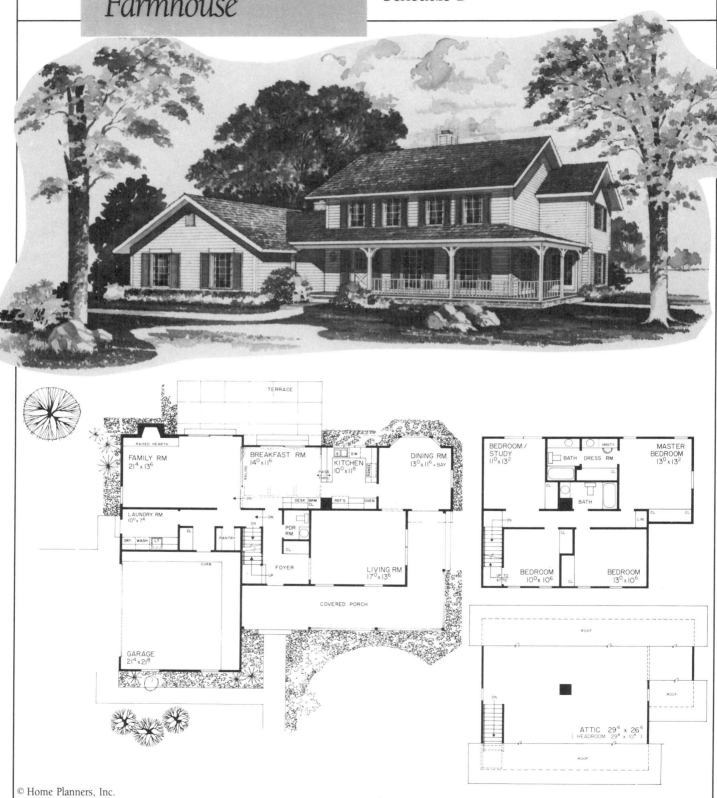

RM-2774
Cost Estimate

Cost at a Glance

Cost per Square Foot: $75.10
Total Cost: $175,658

Cost by Category

		Cost per Square Foot of Living Area		
		Materials	Installation	Total
1. Site Work	Excavation for the basement and footings.		.75	.75
2. Foundation	Main house—12″ concrete block wall on a 20″ x 10″ reinforced concrete footing. Garage—8″ x 42″ reinforced concrete foundation wall. Slabs—4″ thick reinforced steel trowel finished concrete on 4″ compacted gravel.	3.29	4.23	7.52
3. Framing	Main house—2 x 4 studs, 16″ on center with 1/2″ plywood sheathing. Garage—2 x 4 studs, 16″ on center. Floor—2 x 10 joists, 16″ on center with 3/4″ tongue and groove plywood subfloor. Roof—site cut 2 x 6 and 2 x 8 rafters with 1/2″ plywood sheathing.	8.87	5.52	14.39
4. Exterior Walls	Horizontal bevel siding with 15# felt vapor barrier. Vinyl clad double hung, fixed and casement windows and sliding glass doors, crossbuck entry doors. R-19 and R-11 insulation.	10.04	3.50	13.54
5. Roofing	Heavyweight three tab asphalt roof shingles on 30# felt paper. Aluminum drip edge, flashings, gutters and downspouts.	1.73	1.35	3.08
6. Interiors	Wall finish—one coat primer and one coat paint on 1/2″ or 5/8″ gypsum wallboard. Pine door, window and baseboard moldings, painted or stained one coat. Flooring—75% carpet, 22% vinyl, remaining— ceramic tile and hardwood.	7.67	6.43	14.10
7. Specialties	Hardwood faced, particle board case kitchen cabinets and bath vanities with plastic laminate countertops. Washer, dryer, cooktop with hood, double ovens, dishwasher and refrigerator. One masonry fireplace, covered porch and terrace.	5.00	1.90	6.90
8. Mechanical	Oil fired forced hot air heat with central air conditioning. Two full baths, one 1/2 bath. Double bowl kitchen sink with disposal.	3.94	2.65	6.59
9. Electrical	200 amp service, branch circuit wiring with romex cable. Exterior and interior lighting fixtures, receptacles and switches.	2.09	1.23	3.32
10. Overhead	Contractor's overhead and profit.	2.98	1.93	4.91
Total Cost per Square Foot		$45.61	$29.49	$75.10

To purchase a full set of Sepias, Bill of Materials and Detailed Costs—turn to page 267.

RM-3564

2 Story Traditional

2041 Square Feet
3 Bedrooms
2½ Baths
Schedule B

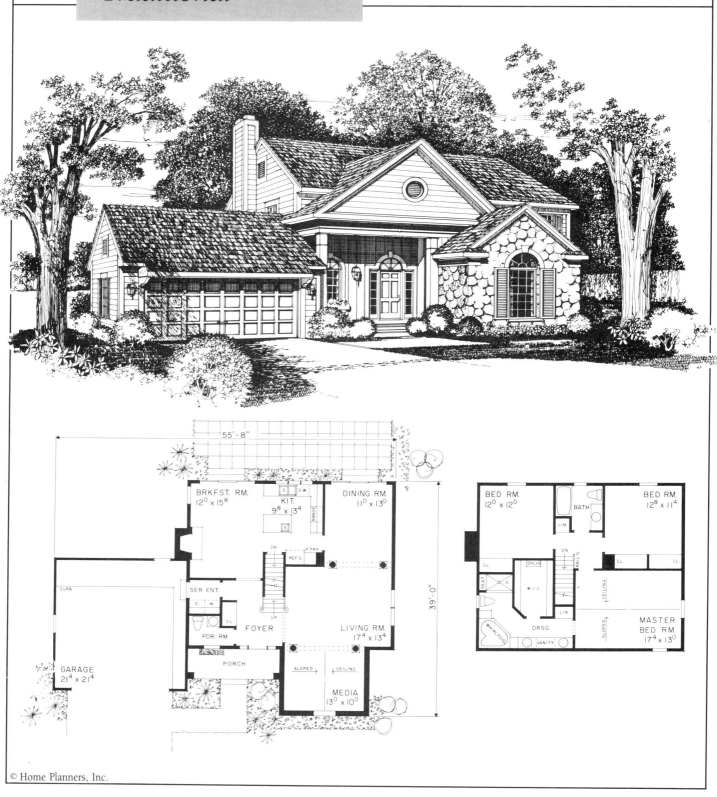

55' - 8"

39' - 0"

BRKFST. RM. 12⁰ x 15⁸

KIT. 9⁸ x 13⁴

DINING RM. 11⁰ x 13⁰

S DW
RANGE
S

DN
P.TRY.
REF'G.

SER. ENT.
D W
CL.

PDR. RM.

FOYER
UP

LIVING RM. 17⁴ x 13⁴

PORCH

SLOPED   CEILING

MEDIA 13⁰ x 10⁰

CURB

GARAGE 21⁴ x 21⁴

BED RM. 12⁰ x 12⁰

BATH

LIN.

BED RM. 12⁸ x 11⁴

CL.

SHELVS
S
W I C
DN
RAIL'G

SEAT

LIN.

WHIRLPOOL
DRSG.
VANITY

CL.
CL.
CEILING

SLOPED

MASTER BED RM. 17⁴ x 13⁰

RM-3564
Cost Estimate

Cost at a Glance
Cost per Square Foot: $80.20
Total Cost: $163,688

Cost by Category

		Cost per Square Foot of Living Area		
		Materials	Installation	Total
1. Site Work	Excavation for the basement and footings.		.77	.77
2. Foundation	Main house — 10″ wide reinforced concrete foundation wall on 20″ x 10″ reinforced concrete perimeter footings. Trench footings — 10″ wide reinforced concrete. Slabs — 4″ thick steel trowel finished reinforced concrete over compacted gravel.	3.17	4.00	7.17
3. Framing	Exterior walls — 2 x 6 studs, 16″ on center with 1/2″ plywood sheathing. Garage — 2 x 4 studs, 16″ on center. Roof — pre-engineered trusses and site cut rafters with 1/2″ plywood sheathing.	8.68	5.29	13.97
4. Exterior Walls	Vertical board, horizontal beveled cedar and stone veneer siding over 15# felt vapor barrier with R-19 and R-11 insulation. Vinyl clad double hung, fixed and casement windows and sliding glass patio doors.	11.67	5.24	16.91
5. Roofing	Heavyweight three tab asphalt shingles over 30# felt roofing paper. Aluminum gutters, downspouts, drip edge and flashings.	1.26	1.04	2.30
6. Interiors	Walls and ceilings — 1/2″ and 5/8″ taped and finished gypsum wallboard, primed and painted with one coat latex. Pine interior trim with one coat paint or stain. Flooring — 73% carpet, 21% vinyl, 4% hardwood and 2% ceramic tile.	7.11	5.90	13.01
7. Specialties	Hardwood faced particle board case kitchen cabinets and bathroom vanities with plastic laminate countertops. Washer, dryer, range with hood, dishwasher and refrigerator. One pre-fabricated fireplace.	7.57	1.54	9.11
8. Mechanical	Oil fired forced hot air heat with central air conditioning. One full bath, one 1/2 bath and a master suite with a whirlpool and shower. Stainless steel kitchen sink with disposal.	5.17	3.16	8.33
9. Electrical	200 amp service, branch circuit wiring with romex cable. Exterior and interior lighting fixtures, receptacles and switches.	2.17	1.21	3.38
10. Overhead	Contractor's overhead and profit.	3.28	1.97	5.25
	Total Cost per Square Foot	$50.08	$30.12	$80.20

To purchase a full set of Sepias, Bill of Materials and Detailed Costs — turn to page 267.

RM-3562

2 Story Transitional

2109 Square Feet
3 Bedrooms
2½ Baths
Schedule B

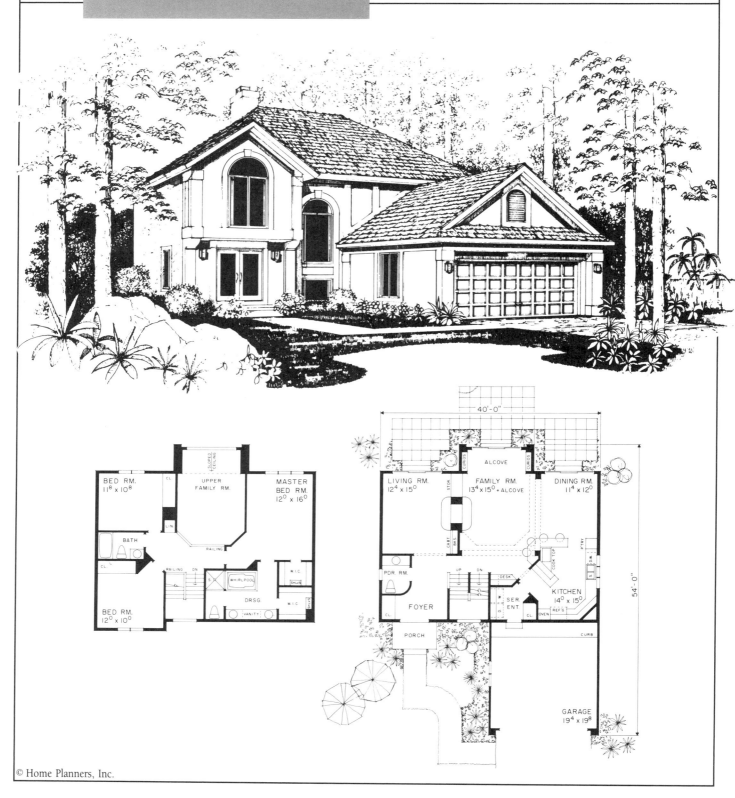

BED RM.
11⁸ x 10⁸

UPPER FAMILY RM.

MASTER BED RM.
12⁰ x 16⁰

BATH

WHIRLPOOL

W.I.C.

BED RM.
12⁰ x 10⁰

DRSG.

VANITY

W.I.C.

RAILING

RAILING DN

LIN.

CL.

CL.

SLOPED CEILING

40'-0"

ALCOVE

LIVING RM.
12⁴ x 15⁰

FAMILY RM.
13⁴ x 15⁰ + ALCOVE

DINING RM.
11⁴ x 12⁰

PDR. RM.

FOYER

UP DN

DESK

SER. ENT.

COOK TOP

KITCHEN
14⁰ x 15⁰

REFG

OVEN

PORCH

CURB

54'-0"

GARAGE
19⁴ x 19⁸

© Home Planners, Inc.

168

RM-3562
Cost Estimate

Cost by Category

Category	Description	Cost per Square Foot of Living Area		
		Materials	Installation	Total
1. Site Work	Excavation for the basement and footings.		.77	.77
2. Foundation	Main house—10″ wide reinforced concrete foundation wall on 20″ x 10″ reinforced concrete perimeter footings. Trench footings—10″ wide reinforced concrete. Slabs—4″ thick steel trowel finished reinforced concrete over compacted gravel.	2.95	3.71	6.66
3. Framing	Exterior walls—2 x 6 studs, 16″ on center with 1/2″ plywood sheathing. Garage—2 x 4 studs, 16″ on center. Roof—pre-engineered trusses and site cut rafters with 1/2″ plywood sheathing.	9.34	6.11	15.45
4. Exterior Walls	Exterior insulation finish system (1½″ thick) siding with R-19 and R-11 insulation. Vinyl clad fixed and casement windows with swinging and sliding glass doors.	7.67	2.36	10.03
5. Roofing	Heavyweight three tab asphalt shingles over 30# felt roofing paper. Aluminum gutters, downspouts, drip edge and flashings.	1.11	1.18	2.29
6. Interiors	Walls and ceilings—1/2″ and 5/8″ taped and finished gypsum wallboard, primed and painted with one coat latex. Pine interior trim with one coat paint or stain. Flooring—74% carpet, 12% vinyl, 8% hardwood and 6% ceramic tile.	6.57	5.99	12.56
7. Specialties	Hardwood faced particle board case kitchen cabinets and bathroom vanities with plastic laminate countertops. Washer, dryer, cooktop with hood, single wall oven, dishwasher and refrigerator. One double faced pre-fabricated fireplace.	10.43	1.88	12.31
8. Mechanical	Oil fired forced hot air heat with central air conditioning. One full bath, one 1/2 bath and a master suite with a whirlpool and shower. Stainless steel kitchen sink with disposal.	4.83	2.97	7.80
9. Electrical	200 amp service, branch circuit wiring with romex cable. Exterior and interior lighting fixtures, receptacles and switches.	2.34	1.26	3.60
10. Overhead	Contractor's overhead and profit.	3.17	1.84	5.01
Total Cost per Square Foot		$48.41	$28.07	$76.48

To purchase a full set of Sepias, Bill of Materials and Detailed Costs—turn to page 267.

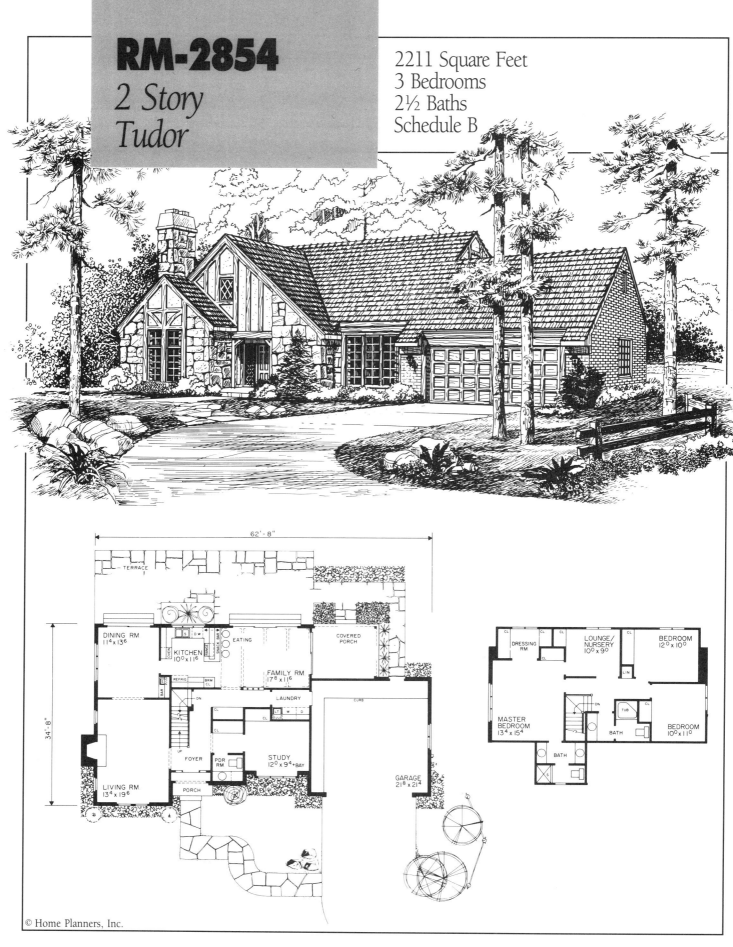

RM-2854

2 Story Tudor

2211 Square Feet
3 Bedrooms
2½ Baths
Schedule B

62'-8"

34'-8"

TERRACE

DINING RM
11⁴ x 13⁶

KITCHEN
10⁰ x 11⁶

EATING

COVERED PORCH

FAMILY RM
17⁸ x 11⁶

REFRIG
BRM CL

LAUNDRY

CL
LT W D
CL

FOYER
PDR RM

STUDY
12⁰ x 9⁴+BAY

CURB

GARAGE
21⁸ x 21⁴

UP
DN

LIVING RM
13⁴ x 19⁶

PORCH

DRESSING RM

LOUNGE/NURSERY
10⁰ x 9⁰

BEDROOM
12⁰ x 10⁰

LIN

MASTER BEDROOM
13⁴ x 15⁴

TUB CL

BATH

BEDROOM
10⁰ x 11⁰

DN

BATH

RM-2854
Cost Estimate

Cost at a Glance

Cost per Square Foot: $85.60
Total Cost: $189,261

Cost by Category

		Cost per Square Foot of Living Area		
		Materials	Installation	Total
1. Site Work	Excavation for the basement and footings.		.75	.75
2. Foundation	Main house—12″ wide concrete masonry unit foundation wall on 20″ x 10″ reinforced concrete perimeter footings. Trench footings—8″ and 12″ wide reinforced concrete. Slabs—4″ thick reinforced steel trowel finished concrete over compacted gravel.	3.65	4.84	8.49
3. Framing	Exterior walls—2 x 6 studs, 16″ on center with 1/2″ plywood sheathing. Garage—2 x 4 studs, 16″ on center. Floors—2 x 10 and 2 x 12 floor joists, 16″ on center with 3/4″ plywood subfloor. Roof— site cut 2 x 12 rafters with 1/2″ plywood sheathing.	10.20	5.80	16.00
4. Exterior Walls	Masonry and stone veneers, and stucco highlights surrounded by wood trim with R-19 and R-11 wall insulation. Vinyl clad fixed, and casement windows and sliding patio doors.	13.21	7.07	20.28
5. Roofing	Heavyweight three tab asphalt shingles over 30# felt roofing paper. Aluminum gutters, downspouts, drip edge and flashings.	1.13	1.11	2.24
6. Interiors	Walls and ceilings—1/2″ and 5/8″ taped and finished gypsum wallboard, primed and painted with one coat latex. Pine interior trim, with one coat paint or stain. Flooring—66% carpet, 18% vinyl, 7% hardwood and 9% ceramic tile.	6.96	5.84	12.80
7. Specialties	Hardwood faced particle board case kitchen cabinets and bathroom vanities with plastic laminate countertops. Washer, dryer, cooktop with hood, double ovens, dishwasher, and refrigerator. One masonry fireplace.	5.18	1.81	6.99
8. Mechanical	Oil fired forced hot air heat with central air conditioning. One full bath one 3/4 bath and one 1/2 bath. Stainless steel double bowl kitchen sink with disposal.	5.18	3.29	8.47
9. Electrical	200 amp service, branch circuit wiring with romex cable. Exterior and interior lighting fixtures, receptacles and switches.	2.51	1.47	3.98
10. Overhead	Contractor's overhead and profit.	3.36	2.24	5.60
Total Cost per Square Foot		$51.38	$34.22	$85.60

To purchase a full set of Sepias, Bill of Materials and Detailed Costs—turn to page 267.

RM-3309

2 Story Victorian

2391 Square Feet
3 Bedrooms
2½ Baths
Schedule B

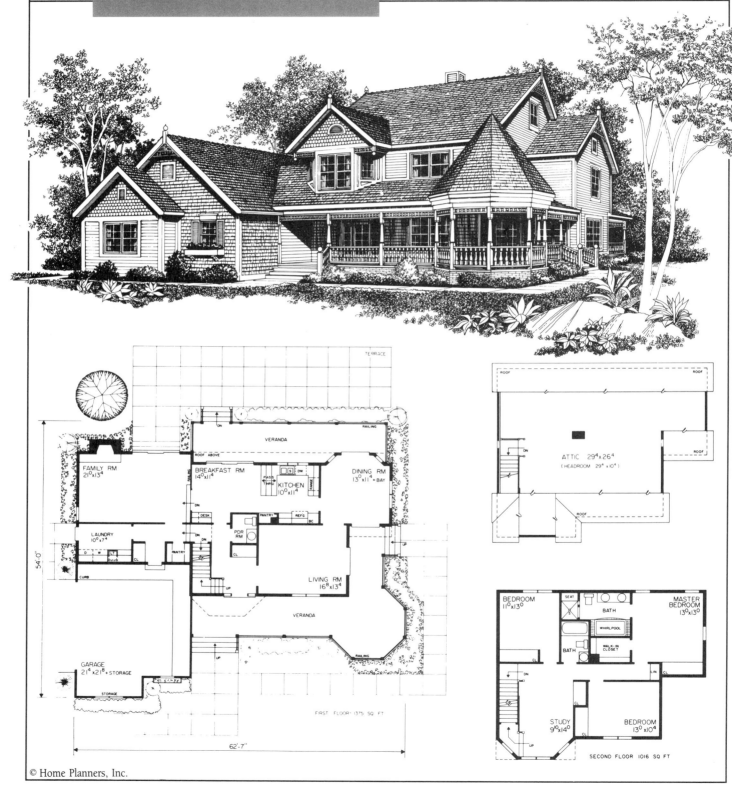

FIRST FLOOR 1375 SQ FT

SECOND FLOOR 1016 SQ FT

© Home Planners, Inc.

RM-3309

Cost Estimate

Cost at a Glance

Cost per Square Foot: $81.95
Total Cost: $195,942

Cost by Category

		Cost per Square Foot of Living Area		
		Materials	Installation	Total
1. Site Work	Excavation for the basement and footings.		.73	.73
2. Foundation	Main house—10″ thick reinforced concrete walls on 16″ and 20″ wide x 10″ high reinforced concrete footings. Garage and porches—8″ and 12″ wide x 42″ high and 8″ x 52″ reinforced concrete trench walls. Slabs—4″ thick reinforced steel trowel finished concrete on 4″ compacted gravel.	3.95	4.78	8.73
3. Framing	Main house—2 x 6 studs, 16″ on center with 1/2″ plywood sheathing. Garage—2 x 4 studs, 16″ on center. Floor—2 x 12 and 2 x 10 joists, 16″ on center with 3/4″ tongue and groove plywood subfloor. Roof—pre-engineered roof trusses and site cut 2 x 6 and 2 x 8 and 2 x 10 rafters, 1/2″ plywood sheathing.	11.60	7.43	19.03
4. Exterior Walls	Horizontal beveled cedar siding, regular and fancy cut shingles over 15# felt vapor barrier. Vinyl clad double hung, casement and fixed windows and sliding glass doors. Glazed French and paneled entry doors. R-19 and R-11 insulation.	11.28	4.34	15.62
5. Roofing	Heavyweight three tab asphalt roof shingles on 30# felt paper. Aluminum drip edge, flashings, gutters and downspouts.	1.82	1.44	3.26
6. Interiors	Wall finish—1/2″ or 5/8″ gypsum wallboard with one coat primer and one coat finish paint. Pine door, window and baseboard moldings. Flooring—70% carpet, 21% vinyl, 5% ceramic tile and 4% hardwood.	6.90	6.66	13.56
7. Specialties	Hardwood faced, particle board case kitchen cabinets and bath vanities with plastic laminate countertops. Washer, dryer, cooktop with hood, double ovens, dishwasher and refrigerator. A masonry fireplace, covered porches and a terrace.	4.35	1.07	5.42
8. Mechanical	Oil fired forced hot air heat with central air conditioning. One full bath, one 1/2 bath and a master suite with a whirlpool tub and shower. Double bowl kitchen sink with disposal.	4.39	2.74	7.13
9. Electrical	200 amp service, branch circuit wiring with romex cable. Exterior and interior lighting fixtures, receptacles and switches.	1.96	1.15	3.11
10. Overhead	Contractor's overhead and profit.	3.24	2.12	5.36
	Total Cost per Square Foot	$49.49	$32.46	$81.95

To purchase a full set of Sepias, Bill of Materials and Detailed Costs—turn to page 267.

RM-1850

Multilevel Traditional

2184 Square Feet
3 Bedrooms
3 Baths
Schedule B

STUDY-BED RM.
11⁰x10⁰

LAUNDRY

CL.
CL.
W
D
BATH
AIR COND.
CL.

UP

CURB

RAISED HEARTH

WOOD BOX

FAMILY RM.
19⁴x14⁰

DN. UP
ENTRY

GARAGE
23⁴x 24⁴

P.

RAILING

DECK

DN.

DINING RM.
11⁰x 12⁰

BREAKFAST
7⁰x 12⁰

RANGE
S.
D.W.
REF'G.

CL.
CL.

DRESS. RM.

MASTER BED RM.
14⁰x 13⁶

BATH

VANITY

KIT.
9⁰x12⁰

BATH

VANITY

CL.

LINEN

PANTRY
DESK
CHINA
G.
CL.

3' HI STORAGE

DN. UP
ENTRY

LIVING RM.
19⁸x15⁰

BED RM.
10⁰x 10⁰

CL.

CL.

BED RM.
11⁰x 13⁶

P.

RM-1850
Cost Estimate

Cost at a Glance

Cost per Square Foot: $71.64
Total Cost: $156,461

Cost by Category

		Cost per Square Foot of Living Area		
		Materials	Installation	Total
1. Site Work	Excavation for the basement and footings.		.39	.39
2. Foundation	Main house — 12″ concrete block on 16″ x 8″ reinforced concrete footings. Garage — 12″ x 42″ reinforced concrete wall. Slabs — 4″ reinforced steel trowel finished concrete over 4″ compacted gravel fill.	2.14	2.83	4.97
3. Framing	Walls — 2 x 6 studs, 16″ on center with 1/2″ plywood sheathing. Floors — 2 x 10 joists, 16″ on center with 3/4″ plywood subfloor. Roof — pre-engineered trusses with 1/2″ plywood sheathing.	6.32	3.92	10.24
4. Exterior Walls	Beveled cedar siding over 15# felt vapor barrier with R-19 wall insulation. Flagstone veneer exterior finish on lower level. Vinyl clad double hung, fixed and casement windows. Wood entry doors, vinyl clad sliding glass doors.	12.23	5.15	17.38
5. Roofing	Heavyweight three tab asphalt shingles over 30# felt roofing paper. Aluminum drip edge, flashings, gutters and downspouts.	.88	.93	1.81
6. Interiors	1/2″ and 5/8″ gypsum wallboard, taped and finished, primed and painted with one coat latex. Pine interior trim with one coat paint or stain. Flooring — 65% carpet, 35% vinyl, with small amounts of ceramic tile and hardwood.	7.24	5.54	12.78
7. Specialties	Hardwood faced, particle board case kitchen cabinets and bath vanities with plastic laminate countertops. Washer, dryer, cooktop with hood, double ovens, dishwasher and refrigerator. 210 square foot sun deck.	6.65	2.65	9.30
8. Mechanical	Oil fired forced hot air with central air conditioning. Two full and one 3/4 bath. Stainless steel kitchen sink with disposal.	4.28	2.86	7.14
9. Electrical	200 amp service, branch circuit wiring with romex cable. Exterior and interior lighting fixtures, receptacles and switches.	1.87	1.08	2.95
10. Overhead	Contractor's overhead and profit.	2.91	1.77	4.68
	Total Cost per Square Foot	$44.52	$27.12	$71.64

To purchase a full set of Sepias, Bill of Materials and Detailed Costs — turn to page 267.

2500 to 3000 Square Feet

Plans	Style	Stories	Total SF	Bedrms	Baths	Page
RM3368	Contemporary	1 Story	2722	3	2½	178
RM2915	Contemporary	1 Story	2758	3	2½	180
RM2880	Traditional	1 Story	2907	3	2½	182
RM3327	Traditional	1 Story	2881	3	2½	184
RM3348	Traditional	1 Story	2549	4	2½	186
RM3559	Transitional	1 Story	2916	3	2½	188
RM3566	Transitional	1½ Story	2537	3	2½	190
RM3558	Transitional	1½ Story	2931	3	2½	192
RM3441	Transitional	1½ Story	2867	4	3½	194
RM3347	Contemporary	2 Story	2674	3	2½	196
RM3396	Farmhouse	2 Story	2776	4	2½	198
RM3398	Farmhouse	2 Story	2821	3	2½	200
RM3325	Farmhouse	2 Story	2707	4	2½	202
RM2946	Farmhouse	2 Story	2925	4	2½	204
RM3432	Southwestern	2 Story	2797	4	3½	206
RM2908	Traditional	2 Story	2580	4	2½	208
RM2855	Tudor	2 Story	2617	4	2½	210
RM2659	Colonial	3 Story	2507	3	2½	212

RM-3368

1 Story
Contemporary

2722 Square Feet
3 Bedrooms
2½ Baths
Schedule C

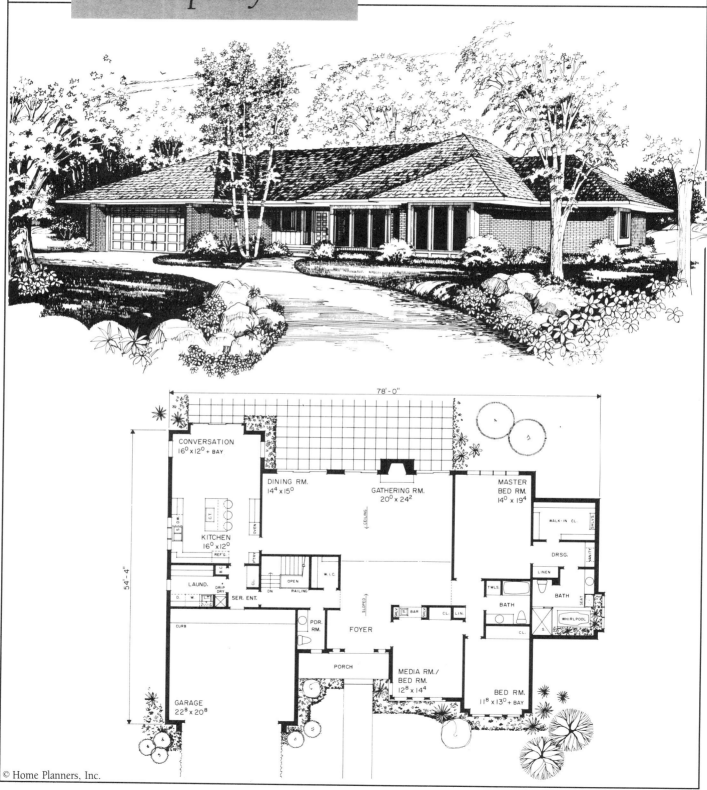

CONVERSATION
16⁰ x 12⁰ + BAY

DINING RM.
14⁴ x 15⁰

GATHERING RM.
20⁰ x 24²

MASTER
BED RM.
14⁰ x 19⁴

WALK-IN CL.

SHLVS.

DRSG.

KITCHEN
16⁰ x 12⁰

LINEN

VANITY

LAUND.

DRIP
DRY

SER. ENT.

TWLS.

BATH

SEAT

BATH

WHIRLPOOL

CURB

PDR.
RM.

FOYER

BAR

CL.

LIN.

GARAGE
22⁸ x 20⁸

PORCH

MEDIA RM./
BED RM.
12⁸ x 14⁴

BED RM.
11⁸ x 13⁰ + BAY

78'-0"

54'-4"

© Home Planners, Inc.

RM-3368

Cost Estimate

Cost by Category

		Cost per Square Foot of Living Area		
		Materials	Installation	Total
1. Site Work	Excavation for the basement and footings.		.96	.96
2. Foundation	Main house — 10″ wide reinforced concrete foundation wall on 20″ x 10″ reinforced concrete perimeter footings. Trench footings — 8″ and 10″ wide reinforced concrete. Slabs — 4″ thick steel trowel finished reinforced concrete over compacted gravel.	3.93	4.12	8.05
3. Framing	Exterior walls — 2 x 4 studs, 16″ on center with 1/2″ plywood sheathing. Floor — 2 x 8 joists, 16″ on center with 3/4″ plywood subfloor. Roof — pre-engineered trusses and site cut rafters with 1/2″ plywood sheathing.	8.03	5.26	13.29
4. Exterior Walls	Brick veneer siding over 15# felt vapor barrier with R-19 and R-11 insulation. Vinyl clad fixed and casement windows and sliding glass patio doors.	9.34	4.28	13.62
5. Roofing	Heavyweight three tab asphalt shingles over 30# felt roofing paper. Aluminum gutters, downspouts, drip edge and flashings.	1.76	1.64	3.40
6. Interiors	Walls and ceilings — 1/2″ and 5/8″ taped and finished gypsum wallboard, primed and painted with one coat latex. Pine interior trim. Flooring — 70% carpet, 24% vinyl, and 6% ceramic tile.	6.37	5.43	11.80
7. Specialties	Hardwood faced particle board case kitchen cabinets and bathroom vanities with plastic laminate countertops. Washer, dryer, cooktop with hood, double wall ovens, dishwasher and refrigerator. One masonry fireplace.	7.91	1.91	9.82
8. Mechanical	Oil fired forced hot air heat with central air conditioning. One full bath, one 1/2 bath and a master suite with a whirlpool and shower. Stainless steel kitchen sink with disposal.	4.95	2.86	7.81
9. Electrical	200 amp service, branch circuit wiring with romex cable. Exterior and interior lighting fixtures, receptacles and switches.	1.92	1.14	3.06
10. Overhead	Contractor's overhead and profit.	3.09	1.93	5.02
	Total Cost per Square Foot	$47.30	$29.53	$76.83

To purchase a full set of Sepias, Bill of Materials and Detailed Costs — turn to page 267.

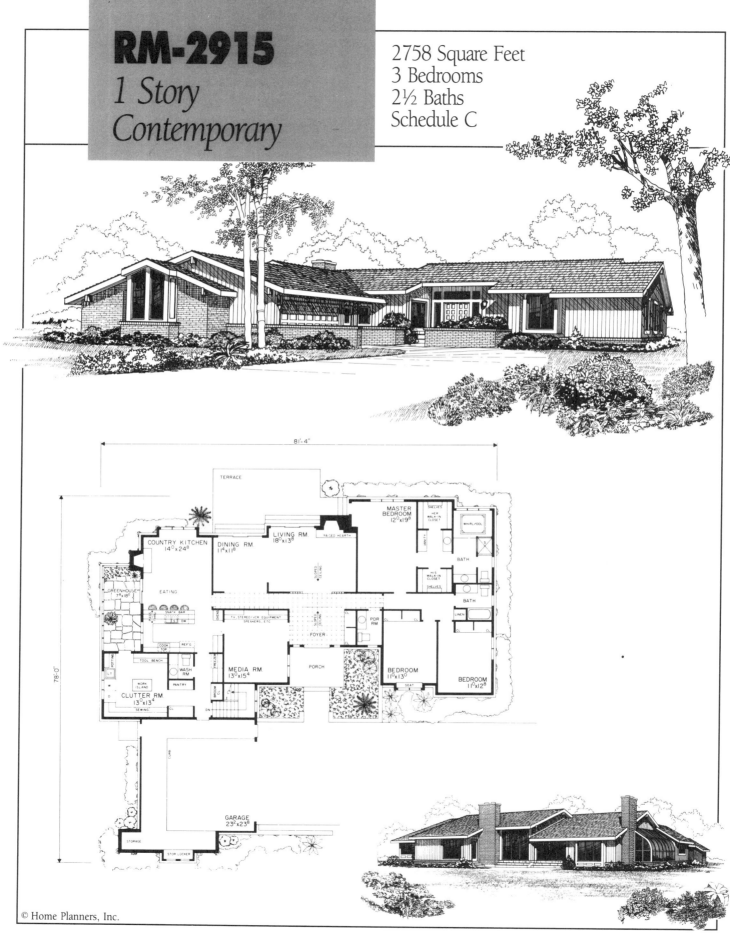

RM-2915

*1 Story
Contemporary*

2758 Square Feet
3 Bedrooms
2½ Baths
Schedule C

81'-4"

78'-0"

TERRACE

COUNTRY KITCHEN
14⁰x24⁸

DINING RM.
11⁴x11⁸

LIVING RM.
18⁰x13⁸

RAISED HEARTH

MASTER BEDROOM
12⁰x19⁸

SHELVES

HER WALK-IN CLOSET

WHIRLPOOL

VANITY

BATH

GREENHOUSE
7⁸x18⁰

EATING

SLOPED CEILING

HIS WALK-IN CLOSET

SHELVES

BATH

SNACK BAR

TV, STEREO/VCR EQUIPMENT SPEAKERS, ETC.

SLOPED CEILING

PDR RM

LINEN

CL

CL

COOK TOP

REF'G

OVEN

FOYER

CL

POTTING

TOOL BENCH

WASH RM

FREEZER

MEDIA RM.
13⁰x15⁴

PORCH

BEDROOM
11⁰x13⁰

SEAT

BEDROOM
11⁰x12⁸

WORK ISLAND

PANTRY

CLUTTER RM.
13⁰x13⁴

SEWING

CL

DN

GARAGE
23²x23⁸

STORAGE

STOR LOCKER

CURB

RM-2915
Cost Estimate

Cost at a Glance

Cost per Square Foot: $92.17
Total Cost: $254,204

Cost by Category

		Cost per Square Foot of Living Area		
		Materials	Installation	Total
1. Site Work	Excavation for the basement and footings.		.96	.96
2. Foundation	Main house — 12″ concrete block wall on a 20″ x 10″ reinforced concrete footing. Trench footings — 8″ and 12″ wide x 42″ reinforced concrete foundation wall. Slabs — 4″ thick reinforced steel trowel finished concrete on 4″ compacted gravel.	4.49	5.89	10.38
3. Framing	Main house — 2 x 6 studs, 16″ on center with 1/2″ plywood sheathing. Garage — 2 x 4 studs, 16″ on center. Floor — 2 x 12 joists, 16″ on center with 3/4″ tongue and groove plywood subfloor. Roof — site cut 2 x 10 rafters with 1/2″ plywood sheathing.	10.81	6.24	17.05
4. Exterior Walls	Vertical pine siding and brick veneer over 15# felt vapor barrier. Vinyl clad fixed and casement windows and sliding glass doors. Paneled and flush entry doors. R-19 and R-11 insulation.	11.99	4.50	16.49
5. Roofing	Heavyweight three tab asphalt roof shingles on 30# felt paper. Aluminum drip edge, flashings, gutters and downspouts.	2.05	1.58	3.63
6. Interiors	Wall finish — one coat primer and one coat paint on 1/2″ or 5/8″ gypsum wallboard. Pine door, window and baseboard moldings. Flooring — 58% carpet, 29% vinyl, 5% ceramic tile and 8% hardwood.	6.44	5.51	11.95
7. Specialties	Hardwood faced, particle board case kitchen cabinets and bath vanities with plastic laminate countertops. Washer, dryer, cooktop with hood, double oven, dishwasher and refrigerator. Two masonry fireplaces, a terrace and a front porch.	11.22	3.11	14.33
8. Mechanical	Oil fired forced hot air heat with central air conditioning. One full bath, two 1/2 baths and a master suite with a shower and whirlpool. Double bowl kitchen sink with disposal.	4.50	2.85	7.35
9. Electrical	200 amp service, branch circuit wiring with romex cable. Exterior and interior lighting fixtures, receptacles and switches.	2.54	1.46	4.00
10. Overhead	Contractor's overhead and profit.	3.78	2.25	6.03
Total Cost per Square Foot		**$57.82**	**$34.35**	**$92.17**

To purchase a full set of Sepias, Bill of Materials and Detailed Costs — turn to page 267.

RM-2880

1 Story Traditional

2907 Square Feet
3 Bedrooms
2½ Baths
Schedule C

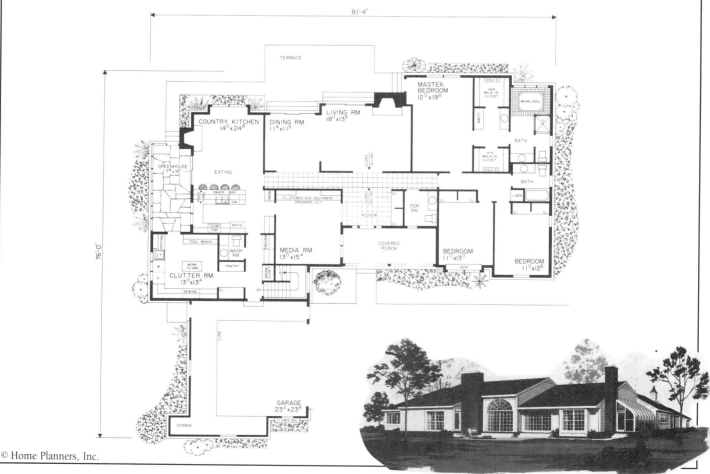

81'-4"

76'-0"

TERRACE

MASTER BEDROOM 12⁰x19⁸

HER WALK-IN CLOSET

SHLVS

WHIRLPOOL

VANITY

BATH

COUNTRY KITCHEN 14⁰x24⁸

DINING RM. 11⁴x11⁸

LIVING RM. 18⁰x13⁸

GREENHOUSE 7⁸x18⁰

EATING

HIS WALK-IN CLOSET

SHLVS

BATH

SNACK BAR

LINEN

WASH DW

COOK TOP

REF'G.

TV, STEREO/VCR EQUIPMENT SPEAKERS, ECT.

FOYER

PDR RM

CL

TOOL BENCH

WASH RM

FREEZER

BEDROOM 11⁰x13⁰

BEDROOM 11⁰x12⁸

WORK ISLAND

PANTRY

COVERED PORCH

SEAT

CLUTTER RM. 13⁰x13⁴

MEDIA RM 13⁰x15⁴

SEWING

CURB

GARAGE 23²x23⁸

STORAGE

RM-2880
Cost Estimate

Cost at a Glance
Cost per Square Foot: $79.54
Total Cost: $231,222

Cost by Category

		Cost per Square Foot of Living Area		
		Materials	Installation	Total
1. Site Work	Excavation for the basement and footings.		.94	.94
2. Foundation	Main house—12″ concrete block wall on a 20″ x 10″ reinforced concrete footing. Garage—12″ x 42″ reinforced concrete foundation wall. Slabs—4″ thick reinforced steel trowel finished concrete on 4″ compacted gravel.	3.26	4.28	7.54
3. Framing	Main house—2 x 6 studs, 16″ on center with 1/2″ plywood sheathing. Garage—2 x 4 studs, 16″ on center. Floor—2 x 12 joists, 16″ on center with 3/4″ tongue and groove plywood subfloor. Roof—site cut 2 x 12 , 2 x 10 and 2 x 8 rafters with 1/2″ plywood sheathing.	10.11	5.54	15.65
4. Exterior Walls	Beveled red cedar horizontal and 12″ wide vertical cedar siding with 15# felt vapor barrier. Vinyl clad fixed, double hung and casement windows and sliding glass doors. Paneled entry doors. R-19 and R-11 insulation.	10.62	4.87	15.49
5. Roofing	Heavyweight three tab asphalt roof shingles on 30# felt paper. Aluminum drip edge, flashings, gutters and downspouts.	1.82	1.41	3.23
6. Interiors	Wall finish—one coat primer and one coat paint on 1/2″ or 5/8″ gypsum wallboard. Pine door, window and baseboard moldings. Flooring—50% carpet, 28% vinyl, 7% ceramic tile and 9% hardwood.	6.06	5.16	11.22
7. Specialties	Hardwood faced, particle board case kitchen cabinets and bath vanities with plastic laminate countertops. Washer, dryer, cooktop with hood, double ovens, dishwasher and refrigerator. Masonry fireplace, a covered porch and terrace.	6.76	2.67	9.43
8. Mechanical	Oil fired forced hot air heat with central air conditioning. Two half baths, one full bath and one master suite including a shower and whirlpool. Double bowl kitchen sink with disposal.	4.38	2.78	7.16
9. Electrical	200 amp service, branch circuit wiring with romex cable. Exterior and interior lighting fixtures, receptacles and switches.	2.35	1.32	3.67
10. Overhead	Contractor's overhead and profit.	3.18	2.03	5.21
	Total Cost per Square Foot	**$48.54**	**$31.00**	**$79.54**

To purchase a full set of Sepias, Bill of Materials and Detailed Costs—turn to page 267.

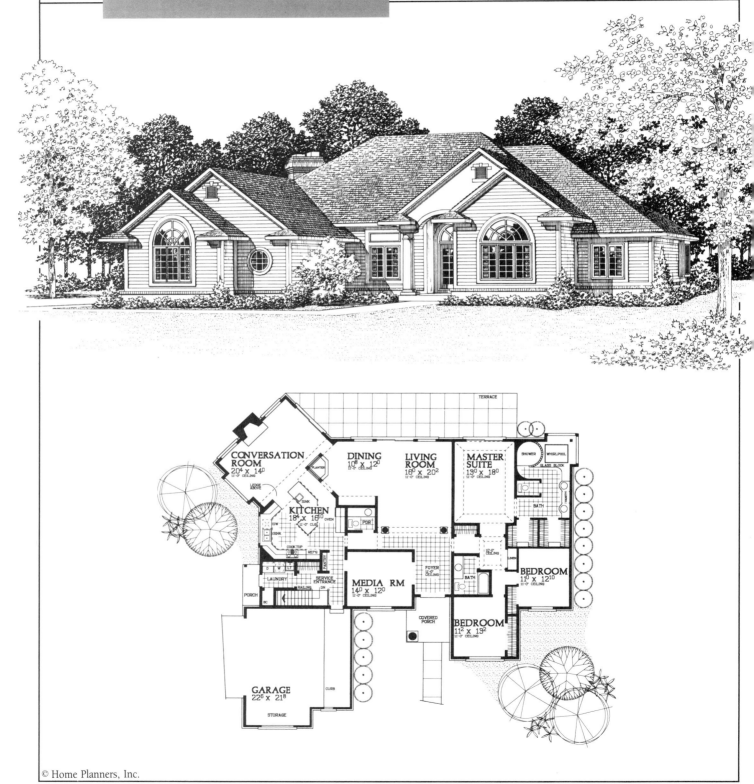

RM-3327

1 Story
Traditional

2881 Square Feet
3 Bedrooms
2½ Baths
Schedule C

TERRACE

CONVERSATION ROOM
20⁴ x 14⁰
11'-0" CEILING

LEDGE ABOVE

PLANTER

DINING
10⁸ x 12⁰
11'-0" CEILING

LIVING ROOM
16⁰ x 20²
11'-0" CEILING

MASTER SUITE
13⁰ x 18⁰
11'-0" CEILING

SHOWER WHIRLPOOL

GLASS BLOCK

BATH

KITCHEN
18⁴ x 16⁰
11'-0" CEILING

SINK

OVEN

PDR

COOK TOP

REF'G

PANTRY

D   W

LAUNDRY

SERVICE ENTRANCE

DN

MEDIA RM
14⁰ x 12⁰
11'-0" CEILING

FOYER
11'-0" CEILING

BATH

LINEN

BEDROOM
11⁰ x 12¹⁰
11'-0" CEILING

PORCH

RAILING

BC

COVERED PORCH

BEDROOM
11² x 13²
11'-0" CEILING

GARAGE
22⁶ x 21⁸

CURB

STORAGE

RM-3327
Cost Estimate

Cost at a Glance

Cost per Square Foot: $86.13
Total Cost: $248,140

Cost by Category

		Cost per Square Foot of Living Area		
		Materials	Installation	Total
1. Site Work	Excavation for the basement and footings.		.95	.95
2. Foundation	Main house—10″ wide reinforced concrete foundation wall on 20″ x 10″ reinforced concrete perimeter footings. Trench footings—10″ wide reinforced concrete. Slabs—4″ thick reinforced steel trowel finished concrete over compacted gravel.	4.10	5.22	9.32
3. Framing	Exterior walls—2 x 6 studs, 16″ on center with 1/2″ plywood sheathing. Garage—2 x 4 studs, 16″ on center. Floors—2 x 10 floor joists, 16″ on center with 3/4″ plywood subfloor. Roof— pre-engineered trusses and site cut rafters with 1/2″ plywood sheathing.	10.77	6.10	16.87
4. Exterior Walls	Beveled cedar siding and MDO plywood over 15# felt vapor barrier with R-19 and R-11 wall insulation. Vinyl clad casement, fixed and half round windows and sliding glass patio doors.	13.52	5.46	18.98
5. Roofing	Heavyweight three tab asphalt shingles over 30# felt roofing paper. Aluminum gutters, downspouts, drip edge and flashings.	2.01	1.74	3.75
6. Interiors	Walls and ceilings—1/2″ and 5/8″ taped and finished gypsum wallboard, primed and painted with one coat latex. Pine interior trim, with one coat paint or stain. Flooring—61% carpet, 16% vinyl, 10% hardwood and 13% ceramic tile.	6.07	5.53	11.60
7. Specialties	Hardwood faced particle board case kitchen cabinets and bathroom vanities with plastic laminate countertops. Washer, dryer, cooktop with hood, double ovens, dishwasher, and refrigerator. One masonry fireplace.	6.82	1.86	8.68
8. Mechanical	Oil fired forced hot air heat with central air conditioning. One full bath and one 1/2 bath and a master suite with a whirlpool and shower. Stainless steel double bowl kitchen sink with disposal.	4.41	2.77	7.18
9. Electrical	200 amp service, branch circuit wiring with romex cable. Exterior and interior lighting fixtures, receptacles and switches.	1.98	1.18	3.16
10. Overhead	Contractor's overhead and profit.	3.48	2.16	5.64
Total Cost per Square Foot		$53.16	$32.97	$86.13

To purchase a full set of Sepias, Bill of Materials and Detailed Costs—turn to page 267.

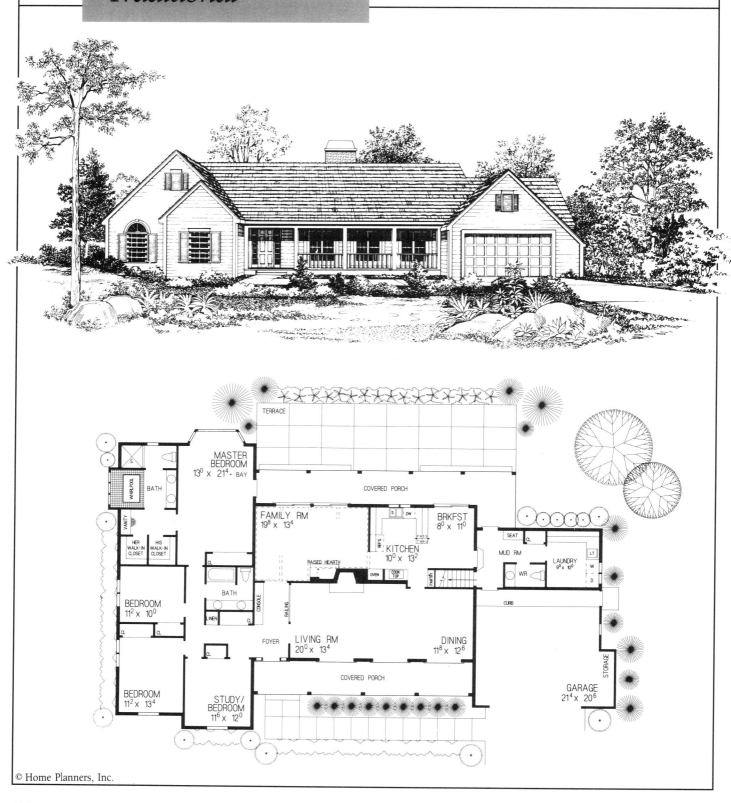

RM-3348

*1 Story
Traditional*

2549 Square Feet
4 Bedrooms
2½ Baths
Schedule C

TERRACE

MASTER BEDROOM
13⁰ x 21⁴ · BAY

COVERED PORCH

WHIRLPOOL

BATH

VANITY

HER WALK-IN CLOSET

HIS WALK-IN CLOSET

FAMILY RM
19⁸ x 13⁴

KITCHEN
10⁰ x 13²

BRKFST
8⁰ x 11⁰

SEAT
CL

MUD RM

LAUNDRY
9⁰ x 10⁰

RAISED HEARTH

OVEN
COOK TOP

PANTRY

WR

W
D

BEDROOM
11² x 10⁰

BATH

LINEN

CONSOLE

RAILING

CURB

FOYER

LIVING RM
20⁰ x 13⁴

DINING
11⁸ x 12⁶

STORAGE

BEDROOM
11² x 13⁴

STUDY/ BEDROOM
11⁶ x 12⁰

COVERED PORCH

GARAGE
21⁴ x 20⁶

RM-3348
Cost Estimate

Cost at a Glance
Cost per Square Foot: $80.05
Total Cost: $204,047

Cost by Category

		Cost per Square Foot of Living Area		
		Materials	Installation	Total
1. Site Work	Excavation for the basement and footings.		.99	.99
2. Foundation	Main house — 10″ wide reinforced concrete foundation wall on 20″ x 10″ reinforced concrete perimeter footings. Trench footings — 8″ and 10″ wide reinforced concrete. Slabs — 4″ thick reinforced steel trowel finished concrete over compacted gravel.	4.52	5.67	10.19
3. Framing	Exterior walls — 2 x 6 studs, 16″ on center with 1/2″ plywood sheathing. Floors — 2 x 10 floor joists, 16″ on center with 3/4″ plywood subfloor. Roof — pre-engineered trusses and site cut rafters with 1/2″ plywood sheathing.	9.57	5.11	14.68
4. Exterior Walls	Beveled cedar siding over 15# felt vapor barrier with R-19 and R-11 wall insulation. Vinyl clad fixed, double hung and casement windows and sliding glass patio doors.	11.51	3.14	14.65
5. Roofing	Heavyweight three tab asphalt shingles over 30# felt roofing paper. Aluminum gutters, downspouts, drip edge and flashings.	1.62	1.32	2.94
6. Interiors	Walls and ceilings — 1/2″ and 5/8″ taped and finished gypsum wallboard, primed and painted with one coat latex. Pine interior trim. Flooring — 51% carpet, 30% vinyl, 8% hardwood and 11% ceramic tile.	6.47	5.81	12.28
7. Specialties	Hardwood faced particle board case kitchen cabinets and bathroom vanities with plastic laminate countertops. Washer, dryer, cooktop with hood, double ovens, dishwasher, and refrigerator. One masonry fireplace.	5.58	2.52	8.10
8. Mechanical	Oil fired forced hot air heat with central air conditioning. One full bath, one 1/2 bath and a master suite with a whirlpool and shower. Stainless steel double bowl kitchen sink with disposal.	4.57	2.89	7.46
9. Electrical	200 amp service, branch circuit wiring with romex cable. Exterior and interior lighting fixtures, receptacles and switches.	2.30	1.22	3.52
10. Overhead	Contractor's overhead and profit.	3.23	2.01	5.24
	Total Cost per Square Foot	$49.37	$30.68	$80.05

To purchase a full set of Sepias, Bill of Materials and Detailed Costs — turn to page 267.

RM-3559

1 Story Transitional

2916 Square Feet
3 Bedrooms
2½ Baths
Schedule C

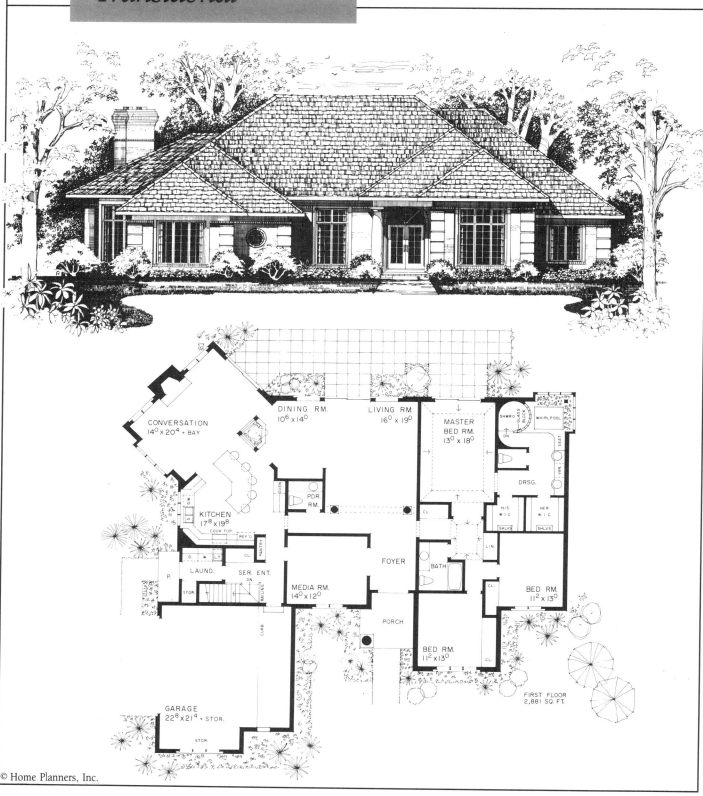

CONVERSATION 14⁰ x 20⁴ + BAY

DINING RM. 10⁶ x 14⁰

LIVING RM. 16⁰ x 19⁰

MASTER BED RM. 13⁰ x 18⁰

SHWR

GLASS BLOCK

WHIRLPOOL

DRSG.

KITCHEN 17⁸ x 19⁸

COOK TOP

REF'G

PANTRY

CL.

HIS W.I.C.

HER W.I.C.

SHLVS

SHLVS

PDR. RM.

OVEN

D   W   LT

LAUND.

SER. ENT.

P.

STOR.

DN

RAILING

MEDIA RM. 14⁰ x 12⁰

FOYER

BATH

LIN.

CL.

BED RM. 11² x 13⁰

PORCH

BED RM. 11² x 13⁰

CL.

CURB

GARAGE 22⁸ x 21⁴ + STOR.

STOR.

FIRST FLOOR 2,881 SQ. FT.

RM-3559
Cost Estimate

Cost at a Glance

Cost per Square Foot: $96.51
Total Cost: $281,423

Cost by Category

		Cost per Square Foot of Living Area		
		Materials	Installation	Total
1. Site Work	Excavation for the basement and footings.		.94	.94
2. Foundation	Main house—10″ wide by 8′ high reinforced concrete foundation wall on 20″ x 10″ reinforced concrete footings. Trench footings—10″ x 42″ reinforced concrete. Slabs—4″ thick reinforced steel trowel finished concrete on 4″ compacted gravel.	6.71	8.34	15.05
3. Framing	Main house and garage—2 x 6 and 2 x 4 studs, 16″ on center with 1/2″ plywood sheathing. Floor—2 x 10 floor joists, 16″ on center with 3/4″ tongue and groove plywood subfloor. Roof—pre-engineered trusses with 5/8″ plywood sheathing.	10.63	6.54	17.17
4. Exterior Walls	Vertical board and exterior plywood siding over 15# felt moisture barrier. Vinyl clad single hung and sliding windows and sliding glass doors. Paneled and flush entry doors. R-19 and R-11 insulation.	13.87	6.53	20.40
5. Roofing	Heavyweight three tab asphalt shingles over 30# felt building paper. Aluminum drip and step flashings, gutters and downspouts.	2.50	1.91	4.41
6. Interiors	Wall finish—1/2″ or 5/8″ gypsum wallboard with one coat primer and one coat finish paint. Pine door, window and baseboard moldings, with one coat paint or stain. Flooring—60% carpet, 19% vinyl, 10% hardwood and 11% ceramic tile.	6.07	5.42	11.49
7. Specialties	Hardwood faced, particle board case kitchen cabinets and bath vanities with plastic laminate countertops. Washer, dryer, cooktop with hood, double ovens, dishwasher and refrigerator. Masonry fireplace.	8.44	2.01	10.45
8. Mechanical	Oil fired forced hot air heat with central air conditioning. One full bath, one 1/2 bath and a master suite with a shower and whirlpool. Double bowl kitchen sink with disposal.	4.39	2.75	7.14
9. Electrical	200 amp service, branch circuit wiring with romex cable. Exterior and interior lighting fixtures, receptacles and switches.	1.98	1.17	3.15
10. Overhead	Contractor's overhead and profit.	3.82	2.49	6.31
Total Cost per Square Foot		$58.41	$38.10	$96.51

To purchase a full set of Sepias, Bill of Materials and Detailed Costs—turn to page 267.

RM-3566

1½ Story Transitional

2537 Square Feet
3 Bedrooms
2½ Baths
Schedule C

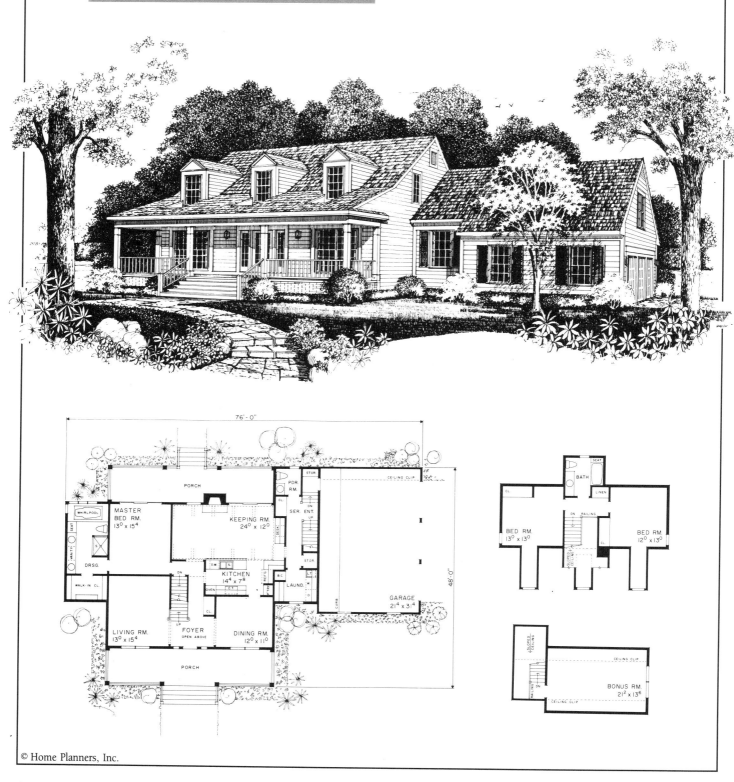

PORCH

MASTER BED RM. 13⁰ x 15⁴

WHIRLPOOL

VANITY SEAT

DRSG.

WALK-IN CL.

KEEPING RM. 24⁰ x 12⁰

PDR. RM.

STOR

SER. ENT.

CEILING CLIP

KITCHEN 14⁴ x 7⁸

OVEN

D.W.

REF'G

B.C.

LAUND.

STOR

GARAGE 21⁴ x 31⁴

LIVING RM. 13⁰ x 15⁴

FOYER OPEN ABOVE

DN UP

DINING RM. 12⁰ x 11⁰

PORCH

76'-0"

48'-0"

CL

BATH

SEAT

LINEN

BED RM. 13⁰ x 13⁰

DN RAILING

SLOPED CEILING

BED RM. 12⁰ x 13⁰

CL

SLOPED CEILING

RAILING DN

CEILING CLIP

BONUS RM. 21² x 13⁶

CEILING CLIP

© Home Planners, Inc.

190

RM-3566
Cost Estimate

Cost at a Glance

Cost per Square Foot: $85.72
Total Cost: $217,471

Cost by Category

		Cost per Square Foot of Living Area		
		Materials	Installation	Total
1. Site Work	Excavation for the basement and footings.		.76	.76
2. Foundation	Main house — 10″ wide reinforced concrete foundation wall on 20″ x 10″ reinforced concrete perimeter footings. Trench footings — 10″ wide reinforced concrete. Slabs — 4″ thick steel trowel finished reinforced concrete over compacted gravel.	4.08	5.00	9.08
3. Framing	Exterior walls — 2 x 6 studs, 16″ on center with 1/2″ plywood sheathing. Garage — 2 x 4 studs, 16″ on center. Roof — site cut rafters with 1/2″ plywood sheathing.	11.90	6.86	18.76
4. Exterior Walls	Beveled cedar siding over 15# felt vapor barrier with R-19 and R-11 insulation. Vinyl clad double hung, fixed and casement windows and sliding glass porch doors.	9.85	3.96	13.81
5. Roofing	Heavyweight three tab asphalt shingles over 30# felt roofing paper. Aluminum gutters, downspouts, drip edge and flashings.	1.55	1.32	2.87
6. Interiors	Walls and ceilings — 1/2″ and 5/8″ taped and finished gypsum wallboard, primed and painted with one coat latex. Pine interior trim with one coat paint or stain. Flooring — 76% carpet, 15% vinyl, 4% hardwood and 5% ceramic tile.	7.19	6.65	13.84
7. Specialties	Hardwood faced particle board case kitchen cabinets and bathroom vanities with plastic laminate countertops. Washer, dryer, cooktop with hood, double ovens, dishwasher and refrigerator. One pre-fabricated fireplace.	8.70	1.67	10.37
8. Mechanical	Oil fired forced hot air heat with central air conditioning. One full bath, one 1/2 bath and a master suite with a whirlpool and shower. Stainless steel kitchen sink with disposal.	4.45	2.79	7.24
9. Electrical	200 amp service, branch circuit wiring with romex cable. Exterior and interior lighting fixtures, receptacles and switches.	2.15	1.23	3.38
10. Overhead	Contractor's overhead and profit.	3.49	2.12	5.61
Total Cost per Square Foot		**$53.36**	**$32.36**	**$85.72**

To purchase a full set of Sepias, Bill of Materials and Detailed Costs — turn to page 267.

RM-3558

1½ Story Transitional

2931 Square Feet
3 Bedrooms
2½ Baths
Schedule C

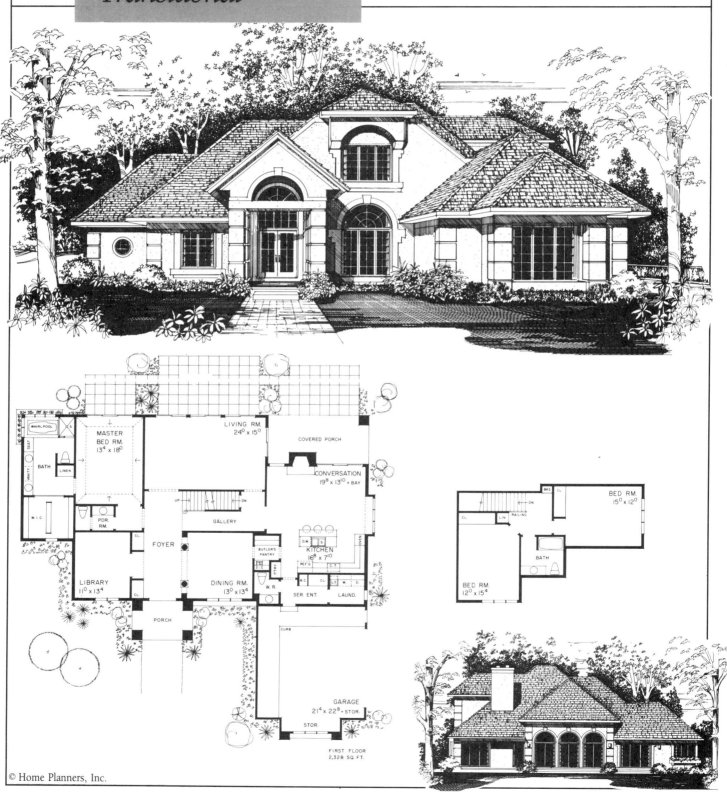

FIRST FLOOR
2,328 SQ. FT.

RM-3558
Cost Estimate

Cost at a Glance

Cost per Square Foot: $78.07
Total Cost: $228,823

Cost by Category

		Cost per Square Foot of Living Area		
		Materials	Installation	Total
1. Site Work	Excavation for the basement and footings.		.81	.81
2. Foundation	Main house — 10″ wide 8′ high reinforced concrete foundation wall on 20″ x 10″ reinforced concrete footings. Trench footings — 10″ x 42″ reinforced concrete. Slabs — 4″ thick reinforced steel trowel finished concrete on 4″ compacted gravel.	3.49	4.42	7.91
3. Framing	Main house — 2 x 6 studs, 16″ on center with 1/2″ plywood sheathing. Garage — 2 x 4 studs, 16″ on center. Floor — 2 x 12 floor joists, 16″ on center with 3/4″ tongue and groove plywood subfloor. Roof — 2 x 8 site cut rafters and pre-engineered trusses with 5/8″ plywood sheathing.	10.57	6.51	17.08
4. Exterior Walls	One coat stucco system on 1½″ foam board siding. Vinyl clad casement and fixed windows and sliding glass doors. Swinging glazed doors, flush mechanical room doors. R-19 and R-11 insulation.	8.94	2.96	11.90
5. Roofing	Heavyweight three tab asphalt shingles over 30# felt building paper. Aluminum drip and step flashings, gutters and downspouts.	2.43	1.91	4.34
6. Interiors	Wall finish — 1/2″ or 5/8″ gypsum wallboard with one coat primer and one coat finish paint. Pine door, window and baseboard moldings, with one coat paint or stain. Flooring — 74% carpet, 11% vinyl, 10% hardwood and 5% ceramic tile.	6.05	5.59	11.64
7. Specialties	Hardwood faced, particle board case kitchen cabinets and bath vanities with plastic laminate countertops. Washer, dryer, cooktop with hood, double ovens, dishwasher and refrigerator. Masonry fireplace, covered porch and patio.	6.74	2.02	8.76
8. Mechanical	Oil fired forced hot air heat with central air conditioning. Two 1/2 baths, a full bath and a master suite with a whirlpool tub and shower. Double bowl kitchen sink with disposal.	4.54	2.88	7.42
9. Electrical	200 amp service, branch circuit wiring with romex cable. Exterior and interior lighting fixtures, receptacles and switches.	1.97	1.13	3.10
10. Overhead	Contractor's overhead and profit.	3.13	1.98	5.11
	Total Cost per Square Foot	$47.86	$30.21	$78.07

To purchase a full set of Sepias, Bill of Materials and Detailed Costs — turn to page 267.

RM-3441
1½ Story Transitional

2867 Square Feet
4 Bedrooms
3½ Baths
Schedule C

63'-8"

56'-2"

PATIO

MASTER BEDROOM
16⁰ X 14⁶ · BAY

SLOPED CEILING

FAMILY RM
18⁰ X 18⁴

SLOPED CEILING

KITCHEN
13⁴ X 19¹⁰

DESK

OVENS

SNACK BAR

COOK TOP

SKYLIGHTS

HIS WALK-IN CLOSET

HER WALK-IN CLOSET

RAILING

PANTRY

REF'S

DW

MASTER BATH

LINEN

BC

D W

LAUNDRY

PDR RM

UP

SLOPED CEILING

DINING RM
11⁴ X 11¹⁰

WHIRLPOOL

LT LINEN

BOOKS

CL

FOYER

SLOPED CEILING

LIVING RM
13⁰ X 15⁰ · BAY

CURB

STUDY
11⁸ X 12⁶

3 CAR GARAGE
31⁴ X 21⁴

OPEN TO FAMILY RM BELOW

PLANT LEDGE

BATH

WALK-IN CLOSET

BEDROOM
13¹⁰ X 11⁰

DN

HALF WALL

LOFT
11⁶ X 6⁰

PLANT LEDGE

OPEN TO FOYER BELOW

BATH

LINEN

SEAT

BEDROOM
11⁰ X 11¹⁰

BEDROOM
10¹⁰ X 12¹⁰

© Home Planners, Inc.

194

RM-3441
Cost Estimate

Cost at a Glance

Cost per Square Foot: $69.78
Total Cost: $200,059

Cost by Category

		Cost per Square Foot of Living Area		
		Materials	Installation	Total
1. Site Work	Excavation for the slab and footings.		.25	.25
2. Foundation	Main house — 6" x 18" and 8" x 18" reinforced concrete walls on 16" x 10" reinforced concrete footings. Spread footings and continuous footings at interior bearing points. Slabs — 4" thick reinforced steel trowel finished concrete on 4" compacted gravel.	1.87	2.08	3.95
3. Framing	Main house and garage — 2 x 6 studs, 16" on center with 1/2" plywood sheathing. Floor — 2 x 12 floor joists, 16" on center with 3/4" tongue and groove plywood subfloor. Roof — pre-engineered trusses, 5/8" plywood sheathing.	7.62	5.61	13.23
4. Exterior Walls	Three coat stucco on high ribbed metal lath siding. Vinyl clad casement, double hung and fixed windows. Paneled, flush and swinging glazed entry doors. R-19 and R-11 insulation.	10.27	1.76	12.03
5. Roofing	Tile roofing over 30# felt building paper. Aluminum valley and step flashings.	6.26	1.62	7.88
6. Interiors	Wall finish — 1/2" or 5/8" gypsum wallboard with one coat primer and one coat finish paint. Pine door, window and baseboard moldings. Flooring — 58% carpet and 42% ceramic tile.	5.46	5.60	11.06
7. Specialties	Hardwood faced, particle board case kitchen cabinets and bath vanities with plastic laminate countertops. Washer, dryer, cooktop, double ovens, dishwasher and refrigerator. Fireplace and patio.	5.01	1.27	6.28
8. Mechanical	Oil fired forced hot air heat with central air conditioning. Two full and one 1/2 bath and a master suite with a whirlpool and shower. Double bowl kitchen sink with disposal.	4.55	2.89	7.44
9. Electrical	200 amp service, branch circuit wiring with romex cable. Exterior and interior lighting fixtures, receptacles and switches.	2.03	1.06	3.09
10. Overhead	Contractor's overhead and profit.	3.02	1.55	4.57
	Total Cost per Square Foot	**$46.09**	**$23.69**	**$69.78**

To purchase a full set of Sepias, Bill of Materials and Detailed Costs — turn to page 267.

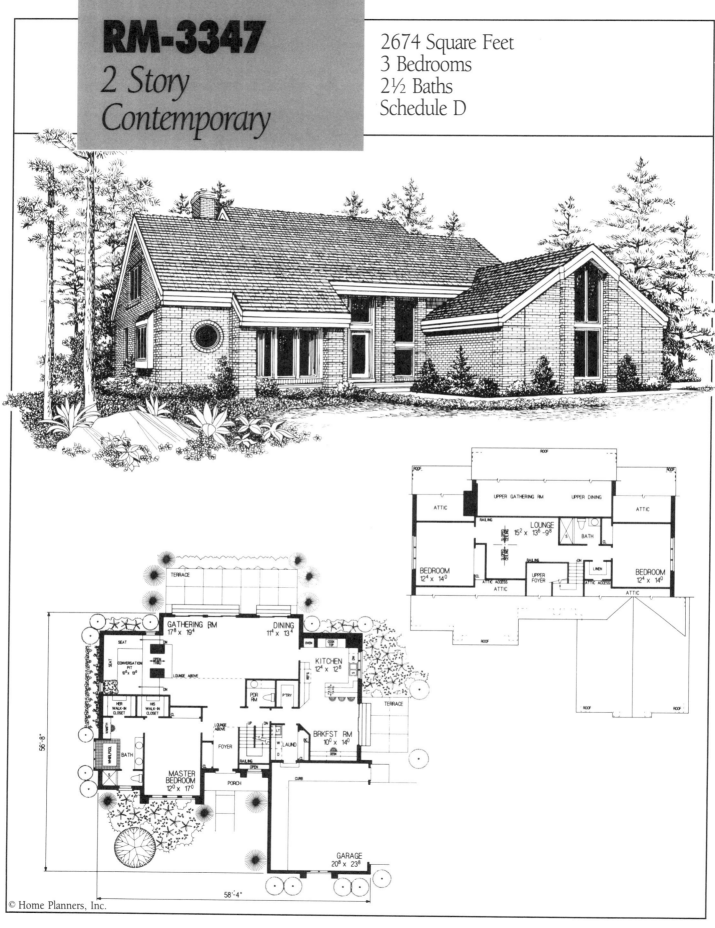

RM-3347

2 Story Contemporary

2674 Square Feet
3 Bedrooms
2½ Baths
Schedule D

GATHERING RM 17⁸ x 19⁴

DINING 11⁴ x 13⁴

TERRACE

CONVERSATION PIT 10⁸ x 13⁰

LOUNGE ABOVE

KITCHEN 12⁴ x 12⁸

HER WALK-IN CLOSET

HIS WALK-IN CLOSET

PDR RM

P'TRY

TERRACE

WHIRLPOOL

BATH

VANITY

LOUNGE ABOVE

FOYER

BRKFST RM 10⁰ x 14⁰

UP DN

LT

W LAUND D

RAILING

OPEN

MASTER BEDROOM 12⁰ x 17⁰

PORCH

CURB

56'-8"

GARAGE 20⁸ x 23⁸

58'-4"

ROOF

ROOF

UPPER GATHERING RM

UPPER DINING

ATTIC

ATTIC

RAILING

LOUNGE 15² x 13⁶ -9⁸

BATH

BEDROOM 12⁴ x 14⁰

RAILING

LINEN

BEDROOM 12⁴ x 14⁰

ATTIC ACCESS

UPPER FOYER

DN

ATTIC ACCESS

ATTIC

ATTIC

ROOF

ROOF

ROOF

© Home Planners, Inc.

RM-3347
Cost Estimate

Cost at a Glance
Cost per Square Foot: $83.08
Total Cost: $222,155

Cost by Category

		Cost per Square Foot of Living Area		
		Materials	Installation	Total
1. Site Work	Excavation for the basement and footings.		.27	.27
2. Foundation	Main house — 10″ wide reinforced concrete foundation wall on 20″ x 10″ reinforced concrete perimeter footings. Trench footings — 8″ and 10″ wide reinforced concrete. Slabs — 4″ thick steel trowel finished reinforced concrete over compacted gravel.	3.62	4.50	8.12
3. Framing	Exterior walls — 2 x 6 studs, 16″ on center with 1/2″ plywood sheathing. Garage — 2 x 4 studs, 16″ on center. Floor — 2 x 12 joists, 16″ on center with 3/4″ plywood subfloor. Roof — pre-engineered trusses and site cut rafters with 1/2″ plywood sheathing.	10.70	5.74	16.44
4. Exterior Walls	Brick veneer and beveled cedar siding over 15# felt vapor barrier with R-19 and R-11 insulation. Vinyl clad fixed and casement windows, and sliding glass patio doors.	12.09	6.66	18.75
5. Roofing	Heavyweight three tab asphalt shingles over 30# felt roofing paper. Aluminum gutters, downspouts, drip edge and flashings.	1.17	.96	2.13
6. Interiors	Walls and ceilings — 1/2″ and 5/8″ taped and finished gypsum wallboard, primed and painted with one coat latex. Pine interior trim. Flooring — 69% carpet, 18% vinyl, 5% hardwood and 8% ceramic tile.	6.78	6.29	13.07
7. Specialties	Hardwood faced particle board case kitchen cabinets and bathroom vanities with plastic laminate countertops. Washer, dryer, cooktop with hood, double wall ovens, dishwasher and refrigerator. One double faced masonry fireplace.	5.82	2.69	8.51
8. Mechanical	Oil fired forced hot air heat with central air conditioning. One 3/4 bath, one 1/2 bath and a master suite with a whirlpool and shower. Stainless steel kitchen sink with disposal.	4.31	2.73	7.04
9. Electrical	200 amp service, branch circuit wiring with romex cable. Exterior and interior lighting fixtures, receptacles and switches.	2.13	1.19	3.32
10. Overhead	Contractor's overhead and profit.	3.26	2.17	5.43
	Total Cost per Square Foot	$49.88	$33.20	$83.08

To purchase a full set of Sepias, Bill of Materials and Detailed Costs — turn to page 267.

RM-3396

2 Story Farmhouse

2776 Square Feet
4 Bedrooms
2½ Baths
Schedule C

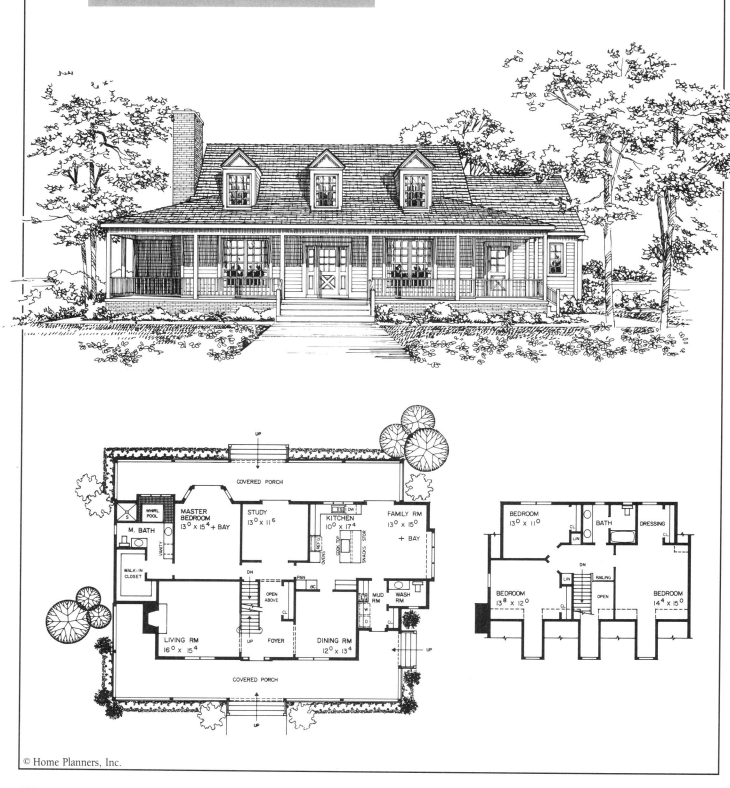

RM-3396
Cost Estimate

Cost at a Glance

Cost per Square Foot: $76.74
Total Cost: $213,030

Cost by Category

		Cost per Square Foot of Living Area		
		Materials	Installation	Total
1. Site Work	Excavation for the basement and footings.		.58	.58
2. Foundation	Main house — 12″ wide masonry foundation wall on 20″ x 10″ reinforced concrete perimeter footings. Trench footings — 8″ and 12″ wide reinforced concrete. Slabs — 4″ thick reinforced steel trowel finished concrete over compacted gravel.	3.47	4.50	7.97
3. Framing	Exterior walls — 2 x 6 studs, 16″ on center with 1/2″ plywood sheathing. Floors — 2 x 12 floor joists, 16″ on center with 3/4″ plywood subfloor. Roof — site cut rafters with 1/2″ plywood sheathing.	10.42	5.93	16.35
4. Exterior Walls	Beveled cedar siding over 15# felt vapor barrier with R-19 wall insulation. Vinyl clad double hung and casement windows and swinging glass patio doors.	9.61	4.12	13.73
5. Roofing	Heavyweight three tab asphalt shingles over 30# felt roofing paper. Aluminum gutters, downspouts, drip edge and flashings.	1.46	1.35	2.81
6. Interiors	Walls and ceilings — 1/2″ and 5/8″ taped and finished gypsum wallboard, primed and painted with one coat latex. Pine interior trim, with one coat paint or stain. Flooring — 69% carpet, 19% vinyl, 6% hardwood and 6% ceramic tile.	6.77	5.75	12.52
7. Specialties	Hardwood faced particle board case kitchen cabinets and bathroom vanities with plastic laminate countertops. Washer, dryer, cooktop with hood, dishwasher, and refrigerator. One masonry fireplace, and two covered porches.	5.56	2.09	7.65
8. Mechanical	Oil fired forced hot air heat with central air conditioning. One full and one 1/2 baths and a master suite with a whirlpool and shower. Stainless steel double bowl kitchen sink with disposal.	4.37	2.76	7.13
9. Electrical	200 amp service, branch circuit wiring with romex cable. Exterior and interior lighting fixtures, receptacles and switches.	1.88	1.10	2.98
10. Overhead	Contractor's overhead and profit.	3.05	1.97	5.02
	Total Cost per Square Foot	$46.59	$30.15	$76.74

To purchase a full set of Sepias, Bill of Materials and Detailed Costs — turn to page 267.

RM-3398

2 Story Farmhouse

2821 Square Feet
3 Bedrooms
2½ Baths
Schedule C

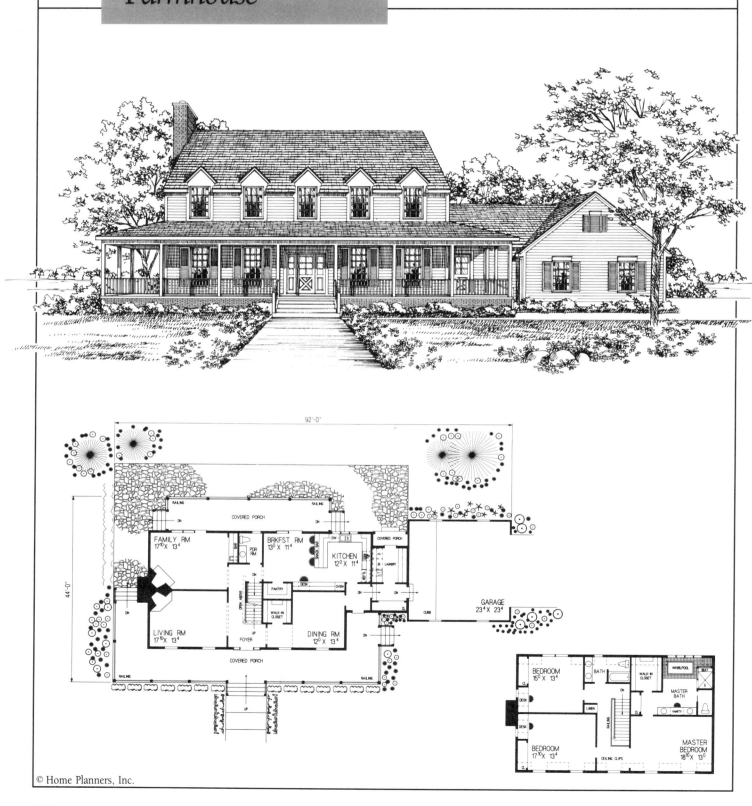

RM-3398
Cost Estimate

Cost at a Glance

Cost per Square Foot: $84.17
Total Cost: $237,443

Cost by Category

		Cost per Square Foot of Living Area		
		Materials	Installation	Total
1. Site Work	Excavation for the basement and footings.		.66	.66
2. Foundation	Main house — 12″ wide concrete masonry unit foundation wall on 20″ x 10″ reinforced concrete perimeter footings. Trench footings — 8″ and 12″ wide reinforced concrete walls. Slabs — 4″ thick steel trowel finished reinforced concrete over compacted gravel.	4.46	5.74	10.20
3. Framing	Exterior walls — 2 x 6 studs, 16″ on center with 1/2″ plywood sheathing. Garage — 2 x 4 studs, 16″ on center. Floor — 2 x 10 joists, 16″ on center with 3/4″ plywood subfloor. Roof — pre-engineered trusses and site cut rafters with 1/2″ plywood sheathing.	9.77	5.89	15.66
4. Exterior Walls	Beveled cedar siding over 15# felt vapor barrier with R-19 and R-11 insulation. Vinyl clad fixed, double hung and casement windows, and swinging glass patio doors.	10.09	4.82	14.91
5. Roofing	Heavyweight three tab asphalt shingles over 30# felt roofing paper. Aluminum gutters, downspouts, drip edge and flashings.	1.58	1.36	2.94
6. Interiors	Walls and ceilings — 1/2″ and 5/8″ taped and finished gypsum wallboard, primed and painted with one coat latex. Pine interior trim with one coat paint or stain. Flooring — 67% carpet, 20% vinyl, 4% hardwood and 9% ceramic tile.	6.74	6.07	12.81
7. Specialties	Hardwood faced particle board case kitchen cabinets and bathroom vanities with plastic laminate countertops. Washer, dryer, cooktop with hood, double ovens, dishwasher and refrigerator. Two masonry fireplaces.	7.65	3.34	10.99
8. Mechanical	Oil fired forced hot air heat with central air conditioning. One full bath, one 1/2 bath and a master suite with a whirlpool and shower. Stainless steel kitchen sink with disposal.	4.37	2.81	7.18
9. Electrical	200 amp service, branch circuit wiring with romex cable. Exterior and interior lighting fixtures, receptacles and switches.	2.16	1.15	3.31
10. Overhead	Contractor's overhead and profit.	3.28	2.23	5.51
	Total Cost per Square Foot	**$50.10**	**$34.07**	**$84.17**

To purchase a full set of Sepias, Bill of Materials and Detailed Costs — turn to page 267.

RM-3325

2 Story Farmhouse

2707 Square Feet
4 Bedrooms
2½ Baths
Schedule C

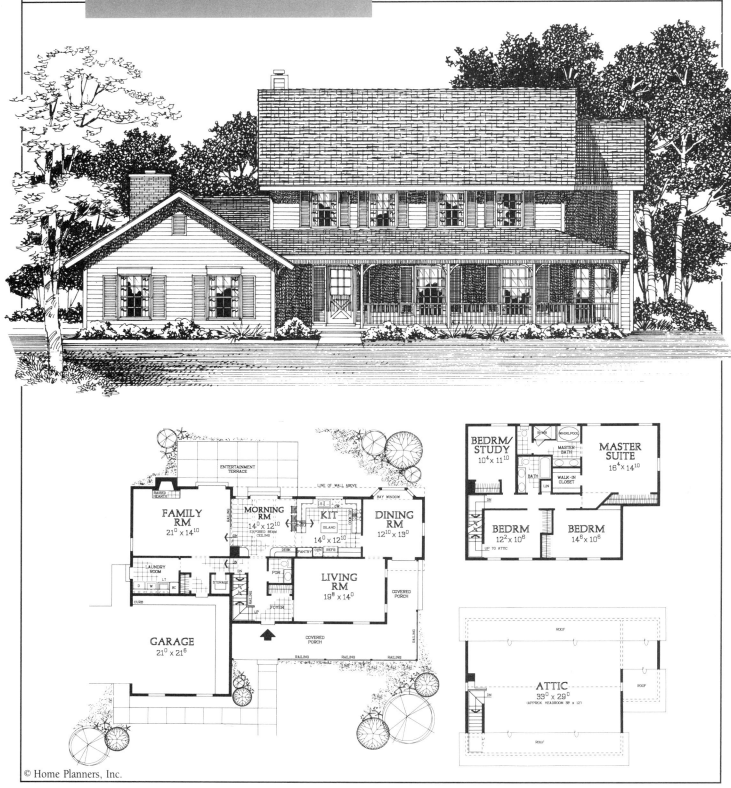

RM-3325
Cost Estimate

Cost by Category

		Cost per Square Foot of Living Area		
		Materials	Installation	Total
1. Site Work	Excavation for the basement and footings.		.70	.70
2. Foundation	Main house — 12″ wide concrete masonry unit foundation wall on 24″ x 10″ reinforced concrete perimeter footings. Trench footings — 8″ and 12″ wide reinforced concrete. Slabs — 4″ thick steel trowel finished reinforced concrete over compacted gravel.	2.97	3.76	6.73
3. Framing	Exterior walls — 2 x 6 studs, 16″ on center with 1/2″ plywood sheathing. Garage — 2 x 4 studs, 16″ on center. Floor — 2 x 10 joists, 16″ on center with 3/4″ plywood subfloor. Roof — pre-engineered trusses and site cut rafters with 1/2″ plywood sheathing.	9.51	6.20	15.71
4. Exterior Walls	Beveled cedar siding over 15# felt vapor barrier with R-19 and R-11 insulation. Vinyl clad fixed, double hung and casement windows with sliding glass patio doors.	11.05	4.33	15.38
5. Roofing	Heavyweight three tab asphalt shingles over 30# felt roofing paper. Aluminum gutters, downspouts, drip edge and flashings.	1.81	1.50	3.31
6. Interiors	Walls and ceilings — 1/2″ and 5/8″ taped and finished gypsum wallboard, primed and painted with one coat latex. Pine interior trim, with one coat paint or stain. Flooring — 66% carpet, 19% vinyl, 7% hardwood and 8% ceramic tile.	6.35	5.60	11.95
7. Specialties	Hardwood faced particle board case kitchen cabinets and bathroom vanities with plastic laminate countertops. Washer, dryer, cooktop with hood, double wall ovens, dishwasher and refrigerator. One masonry fireplace.	5.86	1.99	7.85
8. Mechanical	Oil fired forced hot air heat with central air conditioning. One full bath, one 1/2 bath and a master suite with a whirlpool and shower. Stainless steel kitchen sink with disposal.	3.96	2.69	6.65
9. Electrical	200 amp service, branch circuit wiring with romex cable. Exterior and interior lighting fixtures, receptacles and switches.	2.25	1.34	3.59
10. Overhead	Contractor's overhead and profit.	3.06	1.97	5.03
Total Cost per Square Foot		$46.82	$30.08	$76.90

RM-2946

2 Story Farmhouse

2925 Square Feet
4 Bedrooms
2½ Baths
Schedule C

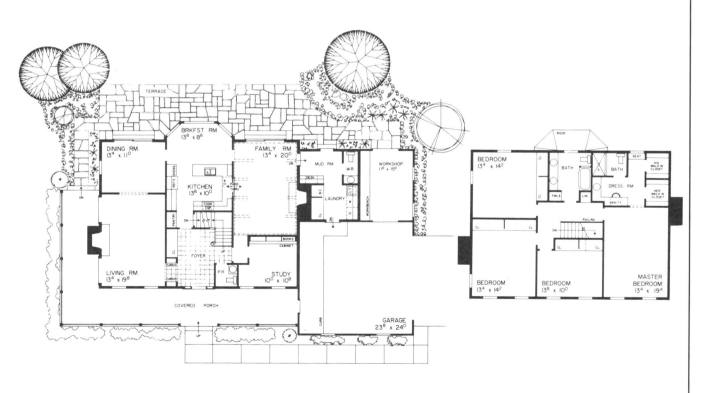

RM-2946
Cost Estimate

Cost at a Glance

Cost per Square Foot: $71.18
Total Cost: $208,201

Cost by Category

		Cost per Square Foot of Living Area		
		Materials	Installation	Total
1. Site Work	Excavation for the basement and footings.		.64	.64
2. Foundation	Main house—10″ thick reinforced concrete walls on 18″ x 10″ reinforced concrete footings. Garage—8″ x 54″ reinforced concrete trench walls. Slabs—4″ thick reinforced steel trowel finished concrete on 4″ compacted gravel.	3.35	3.76	7.11
3. Framing	Main house—2 x 6 studs, 16″ on center with 1/2″ plywood sheathing. Garage—2 x 4 studs, 16″ on center. Floor—2 x 10 joists, 16″ on center with 3/4″ tongue and groove plywood subfloor. Roof— pre-engineered trusses and site cut 2 x 6 rafters with 1/2″ plywood sheathing.	7.97	4.97	12.94
4. Exterior Walls	Bevel cedar siding over 15# felt vapor barrier. Vinyl clad fixed, and double hung windows and sliding glass doors. Crossbuck and paneled entry doors. R-19 and R-11 insulation.	9.19	3.12	12.31
5. Roofing	Heavyweight three tab asphalt roof shingles on 30# felt paper. Aluminum drip edge, flashings, gutters and downspouts.	1.60	1.25	2.85
6. Interiors	Wall finish—one coat primer and one coat paint on 1/2″ or 5/8″ gypsum wallboard. Pine door, window and baseboard moldings. Flooring—76% carpet, 15% vinyl, 5% ceramic tile and 4% hardwood.	6.90	5.88	12.78
7. Specialties	Hardwood faced, particle board case kitchen cabinets and bath vanities with plastic laminate countertops. Washer, dryer, cooktop with hood, double ovens, dishwasher and refrigerator. Two pre-fabricated fireplaces, a covered front porch and a terrace.	6.68	1.74	8.42
8. Mechanical	Oil fired forced hot air heat with central air conditioning. One full bath, one 3/4 bath and two 1/2 baths. Double bowl kitchen sink with disposal.	3.75	2.56	6.31
9. Electrical	200 amp service, branch circuit wiring with romex cable. Exterior and interior lighting fixtures, receptacles and switches.	1.98	1.18	3.16
10. Overhead	Contractor's overhead and profit.	2.90	1.76	4.66
Total Cost per Square Foot		$44.32	$26.86	$71.18

To purchase a full set of Sepias, Bill of Materials and Detailed Costs—turn to page 267.

RM-3432

2 Story Southwestern

2797 Square Feet
4 Bedrooms
3½ Baths
Schedule C

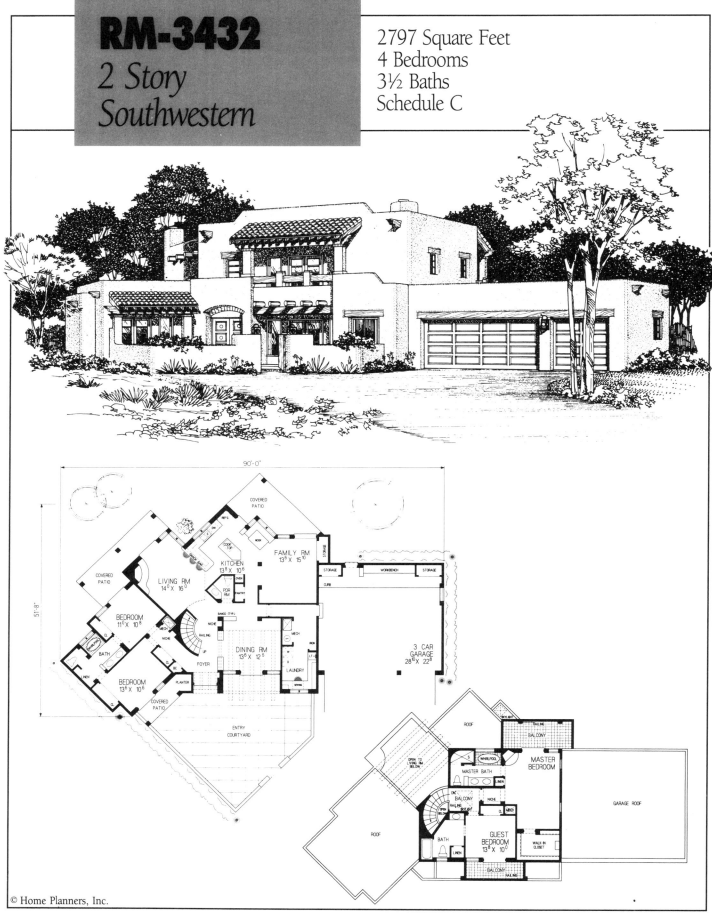

RM-3432
Cost Estimate

Cost at a Glance

Cost per Square Foot: $66.23
Total Cost: $185,245

Cost by Category

		Cost per Square Foot of Living Area		
		Materials	Installation	Total
1. Site Work	Excavation for the slab and footings.		.26	.26
2. Foundation	Main house — 6″ x 18″ reinforced concrete walls on 16″ x 10″ reinforced concrete footings. Spread footings and continuous footings at interior bearing points. Slabs — 4″ thick reinforced steel trowel finished concrete on 4″ compacted gravel.	2.25	2.44	4.69
3. Framing	Main house and garage — 2 x 6 studs, 16″ on center with 1/2″ plywood sheathing. Roof — combination of 10″ diameter wood vigas, engineered lumber, heavy timbers and 2 x 12 rafters with 5/8″ plywood sheathing and 2 x 6 tongue and groove wood decking.	9.69	5.55	15.24
4. Exterior Walls	Three coat stucco on 1″ foam board siding. Vinyl clad casement and fixed windows and swinging glazed doors. Paneled front entry doors. R-19 and R-11 insulation.	7.90	1.78	9.68
5. Roofing	Built-up and mission tile roofing with 30# felt building paper. 4′ x 4′ skylight.	1.51	1.07	2.58
6. Interiors	Wall finish — 1/2″ or 5/8″ gypsum wallboard with one coat primer and one coat finish paint. Pine door, window and baseboard moldings. Flooring — 46% carpet, 46% saltillo floor tile and 8% ceramic tile.	5.52	5.70	11.22
7. Specialties	Hardwood faced, particle board case kitchen cabinets and bath vanities with plastic laminate countertops. Washer, dryer, cooktop, double ovens, microwave ovens, dishwasher and refrigerator. A fireplace and two covered patios.	5.09	1.28	6.37
8. Mechanical	Oil fired forced hot air heat with central air conditioning. One full bath, one 1/2 bath and a guest bath with whirlpool and a master suite with a whirlpool and shower. Double bowl kitchen sink with disposal.	5.05	3.03	8.08
9. Electrical	200 amp service, branch circuit wiring with romex cable. Exterior and interior lighting fixtures, receptacles and switches.	2.56	1.22	3.78
10. Overhead	Contractor's overhead and profit.	2.77	1.56	4.33
Total Cost per Square Foot		$42.34	$23.89	$66.23

To purchase a full set of Sepias, Bill of Materials and Detailed Costs — turn to page 267.

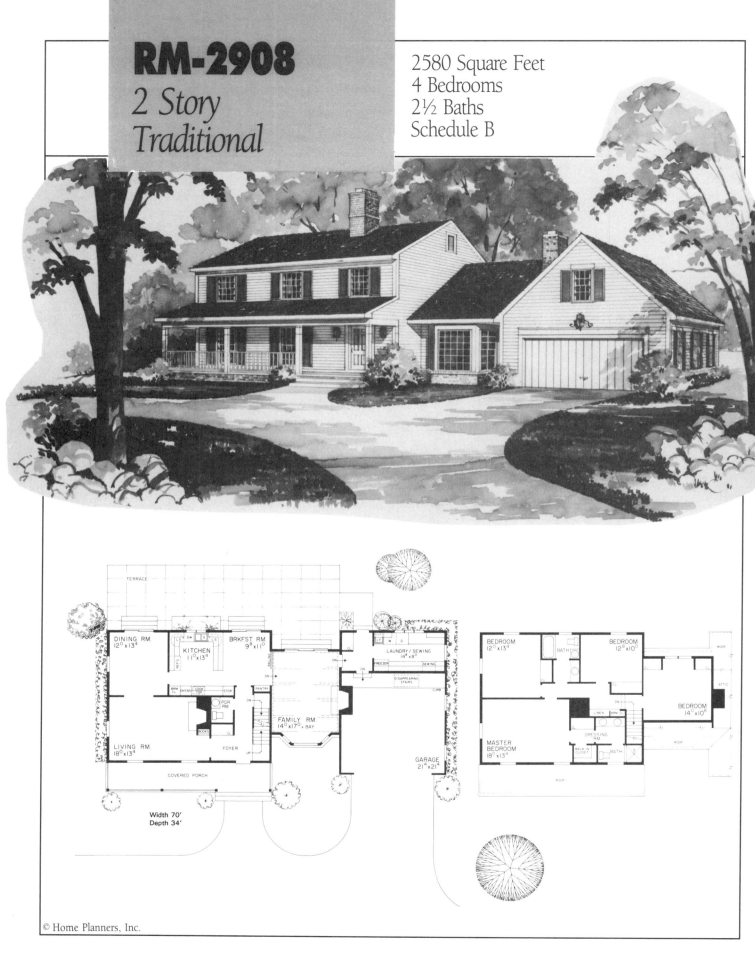

RM-2908

2 Story
Traditional

2580 Square Feet
4 Bedrooms
2½ Baths
Schedule B

TERRACE

DINING RM.
12⁰x13⁴

KITCHEN
11⁰x13⁴

BRKFST RM.
9⁸x11⁰

LAUNDRY / SEWING
14⁸x8⁰

FREEZER   SEWING

DISAPPEARING
STAIRS

CURB

PDR
RM

PANTRY

FAMILY RM.
14⁰x17⁰+ BAY

LIVING RM.
18⁰x13⁴

FOYER

COVERED PORCH

GARAGE
21⁴x21⁴

Width 70'
Depth 34'

BEDROOM
12⁰x13⁴

BATH

BEDROOM
12⁸x10⁰

ROOF

ATTIC

BEDROOM
14⁰x10⁶

LINEN

DRESSING
RM.

MASTER
BEDROOM
18⁰x13⁴

WALK IN
CLOSET

BATH

ROOF

ROOF

RM-2908
Cost Estimate

Cost at a Glance
Cost per Square Foot: $72.19
Total Cost: $186,250

Cost by Category

		Cost per Square Foot of Living Area		
		Materials	Installation	Total
1. Site Work	Excavation for the basement and footings.		.69	.69
2. Foundation	Main house — 12″ concrete block wall on a 20″ x 10″ reinforced concrete footing. Garage and porch — 8″ and 12″ wide x 42″ reinforced concrete foundation wall. Slabs — 4″ thick reinforced steel trowel finished concrete on 4″ compacted gravel.	3.19	4.08	7.27
3. Framing	Main house — 2 x 6 studs, 16″ on center with 1/2″ plywood sheathing. Garage — 2 x 4 studs, 16″ on center. Floor — 2 x 10 and 2 x 12 joists, 16″ on center with 3/4″ tongue and groove plywood subfloor. Roof — site cut 2 x 8 rafters with 1/2″ plywood sheathing.	7.95	4.73	12.68
4. Exterior Walls	Horizontal bevel cedar siding over 15# felt vapor barrier. Vinyl clad fixed, double hung and casement windows and sliding glass doors. Paneled entry doors. R-19 and R-11 insulation.	10.13	3.81	13.94
5. Roofing	Heavyweight three tab asphalt roof shingles on 30# felt paper. Aluminum drip edge, flashings, gutters and downspouts.	1.23	1.08	2.31
6. Interiors	Walls and ceilings — 1/2″ or 5/8″ gypsum primed and painted one coat. Pine door, window and baseboard moldings, one coat paint or stain. Flooring — 70% carpet, 20% vinyl, 5% ceramic tile and 5% hardwood.	6.15	5.25	11.40
7. Specialties	Kitchen cabinets and bathroom vanities — hardwood faced particle board with plastic laminate countertops. Washer, dryer, cooktop with hood, oven, dishwasher and refrigerator. Two masonry fireplaces, a terrace and a front porch.	6.25	2.96	9.21
8. Mechanical	Oil fired forced hot air heat with central air conditioning. One full bath, one 3/4 bath and one 1/2 bath. Double bowl kitchen sink with disposal.	3.87	2.66	6.53
9. Electrical	200 amp service, branch circuit wiring with romex cable. Exterior and interior lighting fixtures, receptacles and switches.	2.13	1.31	3.44
10. Overhead	Contractor's overhead and profit.	2.86	1.86	4.72
Total Cost per Square Foot		**$43.76**	**$28.43**	**$72.19**

To purchase a full set of Sepias, Bill of Materials and Detailed Costs — turn to page 267.

RM-2855

2 Story
Tudor

2617 Square Feet
4 Bedrooms
2½ Baths
Schedule B

RM-2855

Cost Estimate

Cost at a Glance

Cost per Square Foot: $86.92
Total Cost: $227,469

Cost by Category

		Cost per Square Foot of Living Area		
		Materials	Installation	Total
1. Site Work	Excavation for the basement and footings.		.67	.67
2. Foundation	Main house — 12″ concrete block wall on a 20″ x 10″ reinforced concrete footing. Garage — 12″ x 42″ reinforced concrete foundation wall. Slabs — 4″ thick reinforced steel trowel finished concrete on 4″ compacted gravel.	4.17	5.33	9.50
3. Framing	Main house — 2 x 4 studs, 16″ on center with 1/2″ plywood sheathing. Garage — 2 x 4 studs, 16″ on center. Floor — 2 x 10 joists, 16″ on center with 3/4″ tongue and groove plywood subfloor. Roof — pre-engineered roof trusses and site cut 2 x 6, 2 x 8 and 2 x 10 rafters with 1/2″ plywood sheathing.	11.25	6.16	17.41
4. Exterior Walls	Stone, brick, stucco and wood siding. Vinyl clad fixed and casement windows and sliding glass doors. Paneled entry doors. R-19 and R-11 insulation.	13.98	5.85	19.83
5. Roofing	Heavyweight three tab asphalt roof shingles on 30# felt paper. Aluminum drip edge, flashings, gutters and downspouts.	1.72	1.39	3.11
6. Interiors	Wall finish — one coat primer and one coat paint on 1/2″ or 5/8″ gypsum wallboard. Pine door, window and baseboard moldings, painted or stained one coat. Flooring — 65% carpet, 25% vinyl, 5% ceramic tile and 5% hardwood.	6.59	5.08	11.67
7. Specialties	Hardwood faced, particle board case kitchen cabinets and bath vanities with plastic laminate countertops. Washer, dryer, cooktop with hood, double ovens, dishwasher and refrigerator. Two masonry fireplaces, a covered porch and terrace.	6.08	3.02	9.10
8. Mechanical	Oil fired forced hot air heat with central air conditioning. Two full baths, two 1/2 baths. Double bowl kitchen sink with disposal.	4.09	2.80	6.89
9. Electrical	200 amp service, branch circuit wiring with romex cable. Exterior and interior lighting fixtures, receptacles and switches.	1.91	1.15	3.06
10. Overhead	Contractor's overhead and profit.	3.48	2.20	5.68
Total Cost per Square Foot		**$53.27**	**$33.65**	**$86.92**

To purchase a full set of Sepias, Bill of Materials and Detailed Costs — turn to page 267.

RM-2659

3 Story Colonial

2507 Square Feet
3 Bedrooms
2½ Baths
Schedule B

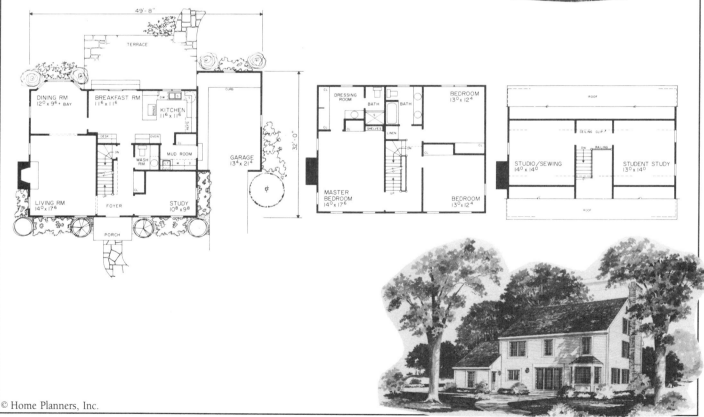

RM-2659
Cost Estimate

Cost at a Glance

Cost per Square Foot: $69.68
Total Cost: $174,687

Cost by Category

		Cost per Square Foot of Living Area		
		Materials	Installation	Total
1. Site Work	Excavation for the basement and footings.		.60	.60
2. Foundation	Main house—12″ wide concrete masonry unit foundation wall on 20″ x 10″ reinforced concrete perimeter footings. Trench footings—8″ wide reinforced concrete. Slabs—4″ thick reinforced steel trowel finished concrete over compacted gravel.	2.37	3.01	5.38
3. Framing	Exterior walls—2 x 4 studs, 16″ on center with 1/2″ plywood sheathing. Floors—2 x 10 floor joists, 16″ on center with 3/4″ plywood subfloor. Roof—pre-engineered trusses and site cut rafters with 1/2″ plywood sheathing.	7.70	4.74	12.44
4. Exterior Walls	Beveled cedar siding over 15# felt vapor barrier with R-13 and R-11 wall insulation. Vinyl clad fixed, casement and double hung windows and sliding patio doors.	9.95	4.29	14.24
5. Roofing	Heavyweight three tab asphalt shingles over 30# felt roofing paper. Aluminum gutters, downspouts, drip edge and flashings.	.87	.75	1.62
6. Interiors	Walls and ceilings—1/2″ and 5/8″ taped and finished gypsum wallboard, primed and painted with one coat latex. Pine interior trim, with one coat paint or stain. Flooring—75% carpet, 18% vinyl, 4% hardwood and 3% ceramic tile.	7.50	6.48	13.98
7. Specialties	Hardwood faced particle board case kitchen cabinets and bathroom vanities with plastic laminate countertops. Washer, dryer, cooktop with hood, double ovens, dishwasher, and refrigerator. One masonry fireplace.	5.10	2.23	7.33
8. Mechanical	Oil fired forced hot air heat with central air conditioning. One full bath, one 3/4 bath and one 1/2 bath. Stainless steel double bowl kitchen sink with disposal.	3.83	2.61	6.44
9. Electrical	200 amp service, branch circuit wiring with romex cable. Exterior and interior lighting fixtures, receptacles and switches.	1.96	1.13	3.09
10. Overhead	Contractor's overhead and profit.	2.75	1.81	4.56
	Total Cost per Square Foot	**$42.03**	**$27.65**	**$69.68**

To purchase a full set of Sepias, Bill of Materials and Detailed Costs—turn to page 267.

3000 to 4500 Square Feet

Plans	Style	Stories	Total SF	Bedrms	Baths	Page
RM3551	Colonial	1½ Story	3076	4	2½	216
RM2920	Contemporary	1½ Story	4011	3	2½	218
RM3428	Southwestern	1½ Story	3174	4	3½	220
RM2921	Traditional	1½ Story	4222	3	2½	222
RM2662	Colonial	2 Story	3556	5	3½	224
RM3399	Farmhouse	2 Story	3818	4	3½	226
RM3414	Southwestern	2 Story	3168	5	3½	228
RM2694	Traditional	2 Story	3412	3	2½	230
RM3334	Traditional	2 Story	3024	4	2½	232
RM2958	Tudor	2 Story	3133	4	2½	234

RM-3551

1½ Story Colonial

3076 Square Feet
4 Bedrooms
2½ Baths
Schedule D

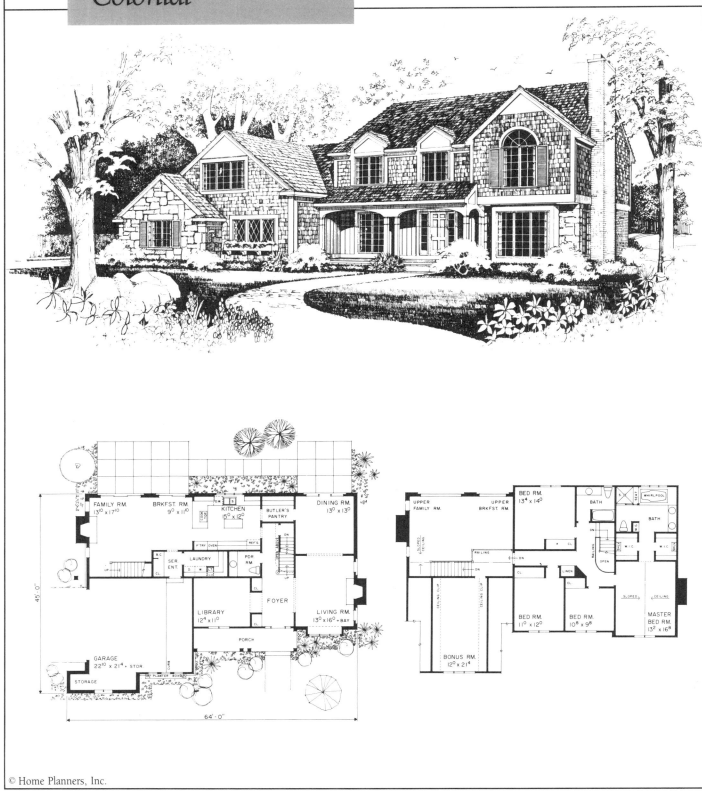

RM-3551

Cost Estimate

Cost at a Glance

Cost per Square Foot: $75.35
Total Cost: $231,776

Cost by Category

		Cost per Square Foot of Living Area		
		Materials	Installation	Total
1. Site Work	Excavation for the basement and footings.		.47	.47
2. Foundation	Main house — 10″ wide reinforced concrete foundation walls on 20″ x 10″ reinforced concrete perimeter footings. Trench footings — 10″ wide reinforced concrete. Slabs — 4″ thick steel trowel finished reinforced concrete over compacted gravel.	2.87	3.68	6.55
3. Framing	Exterior walls — 2 x 6 studs, 16″ on center with 1/2″ plywood sheathing. Garage — 2 x 4 studs, 16″ on center. Floor — 2 x 8 and 2 x 12 joists, 16″ on center with 3/4″ plywood subfloor. Roof — pre-engineered trusses and site cut rafters with 1/2″ plywood sheathing.	9.02	5.61	14.63
4. Exterior Walls	Brick veneer and cedar shingle siding over 15# felt vapor barrier with R-19 and R-11 insulation. Vinyl clad fixed and casement windows with swinging and sliding glass patio doors.	10.57	5.44	16.01
5. Roofing	Heavyweight three tab asphalt shingles over 30# felt roofing paper. Aluminum gutters, downspouts, drip edge and flashings.	1.09	.99	2.08
6. Interiors	Walls and ceilings — 1/2″ and 5/8″ taped and finished gypsum wallboard, primed and painted with one coat latex. Pine interior trim with one coat paint or stain. Flooring — 80% carpet, 11% vinyl, 5% hardwood and 4% ceramic tile.	5.88	5.21	11.09
7. Specialties	Hardwood faced particle board case kitchen cabinets and bathroom vanities with plastic laminate countertops. Washer, dryer, cooktop with hood, double ovens, dishwasher and refrigerator. Two masonry fireplaces.	7.60	2.43	10.03
8. Mechanical	Oil fired forced hot air heat with central air conditioning. One full bath, one 1/2 bath, and a master suite with a whirlpool and shower. Stainless steel kitchen sink with disposal.	4.13	2.59	6.72
9. Electrical	200 amp service, branch circuit wiring with romex cable. Exterior and interior lighting fixtures, receptacles and switches.	1.77	1.08	2.85
10. Overhead	Contractor's overhead and profit.	3.00	1.92	4.92
	Total Cost per Square Foot	**$45.93**	**$29.42**	**$75.35**

To purchase a full set of Sepias, Bill of Materials and Detailed Costs — turn to page 267.

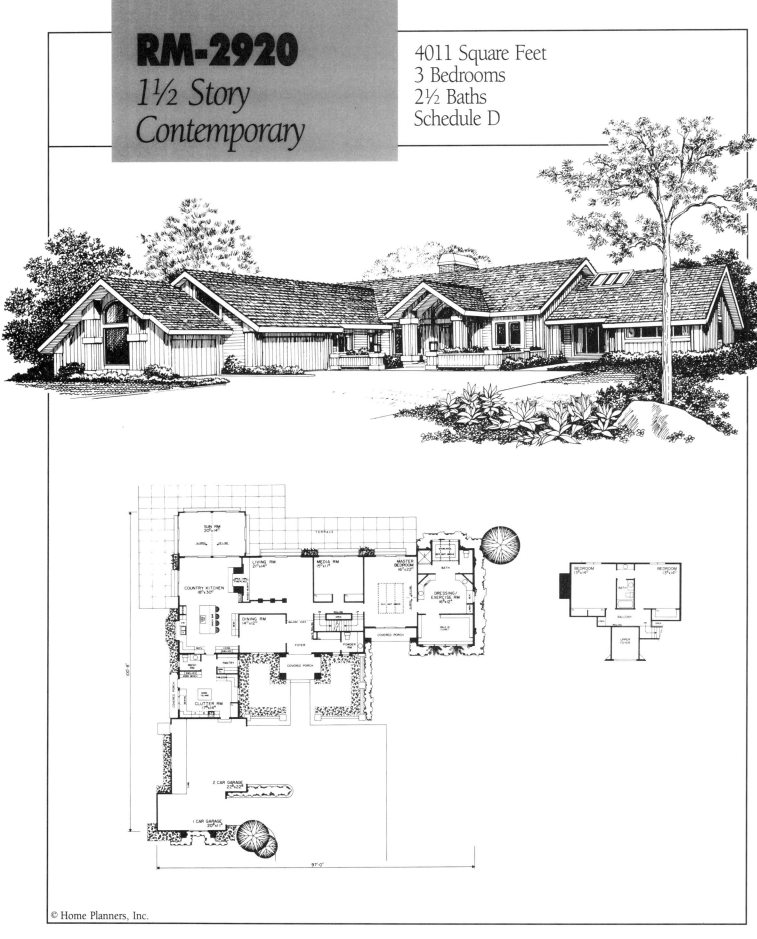

RM-2920

*1½ Story
Contemporary*

4011 Square Feet
3 Bedrooms
2½ Baths
Schedule D

© Home Planners, Inc.

RM-2920
Cost Estimate

Cost at a Glance

Cost per Square Foot: $76.89
Total Cost: $308,405

Cost by Category

		Cost per Square Foot of Living Area		
		Materials	Installation	Total
1. Site Work	Excavation for the basement and footings.		.75	.75
2. Foundation	Main house — 8″ and 12″ concrete block walls on 16″ x 8″ and 20″ x 10″ reinforced concrete footings. Trench footings — 8″ and 12″ wide x 42″ reinforced concrete foundation wall. Slabs — 4″ thick reinforced steel trowel finished concrete on 4″ compacted gravel.	3.05	4.05	7.10
3. Framing	Main house — 2 x 6 studs, 16″ on center with 1/2″ plywood sheathing. Garage — 2 x 4 studs, 16″ on center. Floor — 2 x 12 and 2 x 10 joists, 16″ on center with 3/4″ tongue and groove plywood subfloor. Roof — site cut 2 x 10 rafters with 1/2″ plywood sheathing.	9.25	5.27	14.52
4. Exterior Walls	Horizontal and vertical siding over 15# felt vapor barrier. Vinyl clad fixed and casement windows and sliding glass doors. Paneled and flush entry doors. R-19 and R-11 insulation.	13.59	3.93	17.52
5. Roofing	Heavyweight three tab asphalt roof shingles on 30# felt paper. Aluminum drip edge, flashings, gutters and downspouts. Glazed aluminum skylights.	1.54	1.35	2.89
6. Interiors	Wall finish — one coat primer and one coat paint on 1/2″ or 5/8″ gypsum wallboard. Pine door, window and baseboard moldings, one coat paint or stain. Flooring — 58% carpet, 25% vinyl, 7% ceramic tile and 10% hardwood.	5.90	5.10	11.00
7. Specialties	Hardwood faced, particle board case kitchen cabinets and bath vanities with plastic laminate countertops. Washer, dryer, cooktop with hood, double oven, dishwasher, compactor, freezer and refrigerator. Two faced masonry fireplace and a covered front porch.	6.40	2.07	8.47
8. Mechanical	Oil fired forced hot air heat with central air conditioning. One full bath, two 1/2 baths and a master suite with a shower and whirlpool. Double bowl kitchen sink with disposal.	3.81	2.44	6.25
9. Electrical	200 amp service, branch circuit wiring with romex cable. Exterior and interior lighting fixtures, receptacles and switches.	2.24	1.12	3.36
10. Overhead	Contractor's overhead and profit.	3.20	1.83	5.03
Total Cost per Square Foot		$48.98	$27.91	$76.89

To purchase a full set of Sepias, Bill of Materials and Detailed Costs — turn to page 267.

RM-3428

1½ Story
Southwestern

3174 Square Feet
4 Bedrooms
3½ Baths
Schedule C

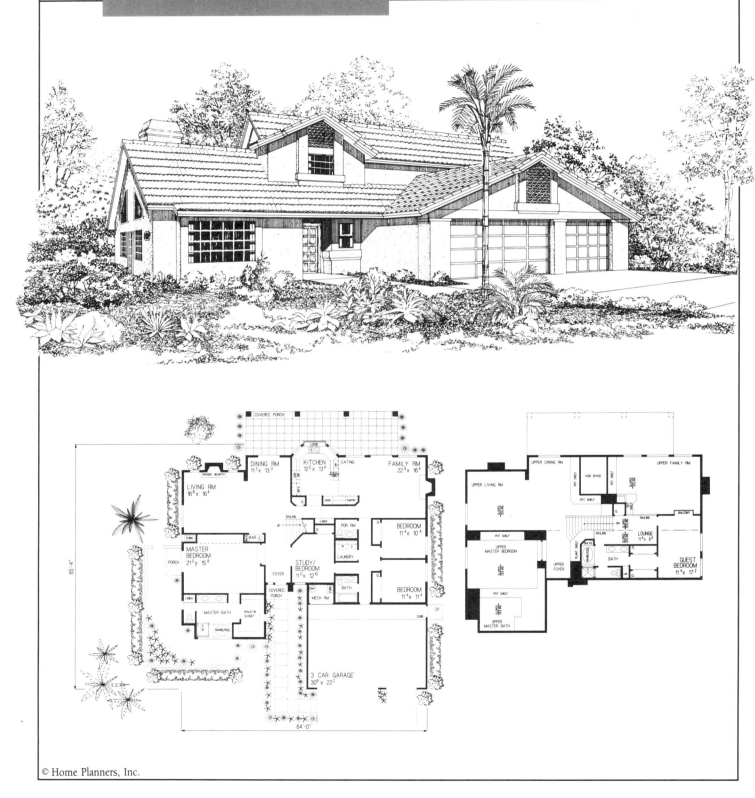

© Home Planners, Inc.

RM-3428
Cost Estimate

Cost at a Glance
Cost per Square Foot: $87.33
Total Cost: $277,185

Cost by Category

		Cost per Square Foot of Living Area		
		Materials	Installation	Total
1. Site Work	Excavation for the slab and footings.		.67	.67
2. Foundation	Main house and garage — 8″ wide reinforced concrete foundation wall on 16″ x 10″ reinforced concrete perimeter footings. Slabs — 4″ thick steel trowel finished reinforced concrete over compacted gravel.	2.66	3.54	6.20
3. Framing	Exterior walls — 2 x 6 studs, 16″ on center with 1/2″ plywood sheathing. Garage — 2 x 4 studs, 16″ on center. Floor — 2 x 10 joists, 16″ on center with 3/4″ plywood subfloor. Roof — pre-engineered trusses and site cut rafters with 5/8″ plywood sheathing.	12.64	13.33	25.97
4. Exterior Walls	3 coat stucco system on high rib metal lath with R-19 and R-11 insulation. Vinyl clad windows and swinging glass patio doors.	6.18	1.43	7.61
5. Roofing	Concrete roof tiles over 30# felt roofing paper. Aluminum gutters, downspouts, drip edge and flashings.	7.51	2.25	9.76
6. Interiors	Walls and ceilings — 1/2″ and 5/8″ taped and finished gypsum wallboard, primed and painted with one coat latex. Pine interior trim with one coat paint or stain. Flooring — 75% carpet, 12% vinyl, 8% hardwood and 5% ceramic tile.	6.37	5.82	12.19
7. Specialties	Hardwood faced particle board case kitchen cabinets and bathroom vanities with plastic laminate countertops. Washer, dryer, cooktop with hood, double ovens, dishwasher and refrigerator. Two pre-manufactured fireplaces.	5.88	1.44	7.32
8. Mechanical	Oil fired forced hot air heat with central air conditioning. One full bath, one 1/2 bath and a master suite and guest suite, both with a whirlpool and shower. Stainless steel kitchen sink with disposal.	5.08	3.04	8.12
9. Electrical	200 amp service, branch circuit wiring with romex cable. Exterior and interior lighting fixtures, receptacles and switches.	2.56	1.22	3.78
10. Overhead	Contractor's overhead and profit.	3.42	2.29	5.71
Total Cost per Square Foot		$52.30	$35.03	$87.33

To purchase a full set of Sepias, Bill of Materials and Detailed Costs — turn to page 267.

RM-2921

1½ Story
Traditional

4222 Square Feet
3 Bedrooms
2½ Baths
Schedule D

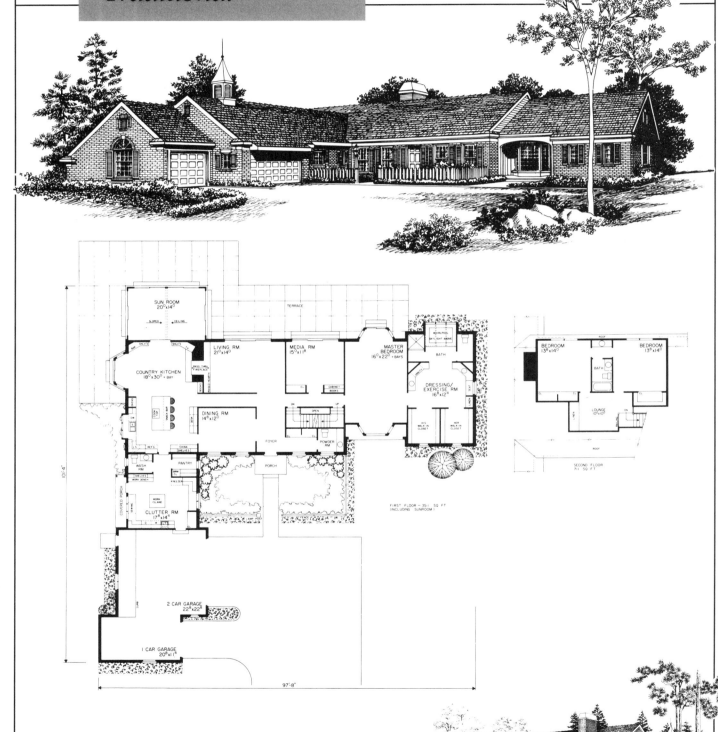

SUN ROOM
20⁰x14⁰

TERRACE

LIVING RM
21⁰x14⁰

MEDIA RM
15⁰x11⁸

MASTER
BEDROOM
16⁰x22⁰ + BAYS

WHIRLPOOL

SKYLIGHT ABOVE

BATH

COUNTRY KITCHEN
18⁰x30⁰ + BAY

PASS THRU
FIREPLACE

CABINET
BOOKS

DRESSING/
EXERCISE RM
16⁸x12⁴

DINING RM
14⁸x12⁰

HIS
WALK IN
CLOSET

HER
WALK IN
CLOSET

FOYER

POWDER
RM

PORCH

PANTRY

WASH
RM

CLUTTER RM
17⁸x14⁴

FIRST FLOOR — 3511 SQ FT
(INCLUDING SUNROOM)

2 CAR GARAGE
22⁸x22⁸

1 CAR GARAGE
20⁸x11⁴

97'-8"

101'-4"

BEDROOM
13⁸x14⁰

BEDROOM
13⁸x14⁰

ROOF

BATH

LOUNGE
10⁰x10⁰

ROOF

SECOND FLOOR
711 SQ FT

RM-2921
Cost Estimate

Cost at a Glance

Cost per Square Foot: $70.64
Total Cost: $298,242

Cost by Category

		Cost per Square Foot of Living Area		
		Materials	Installation	Total
1. Site Work	Excavation for the basement and footings.		.74	.74
2. Foundation	Main house — 8″ and 12″ concrete block walls on 16″ x 8″ and 20″ x 8″ reinforced concrete footings. Trench footings — 8″ and 12″ wide x 42″ reinforced concrete foundation wall. Slabs — 4″ thick reinforced steel trowel finished concrete on 4″ compacted gravel.	3.18	4.20	7.38
3. Framing	Main house — 2 x 6 studs, 16″ on center with 1/2″ plywood sheathing. Garage — 2 x 4 studs, 16″ on center. Floor — 2 x 12 joists, 16″ on center with 3/4″ tongue and groove plywood subfloor. Roof — site cut 2 x 10 rafters with 1/2″ plywood sheathing.	8.92	5.00	13.92
4. Exterior Walls	Brick veneer siding over 15# felt vapor barrier. Vinyl clad fixed, double hung and casement windows and sliding glass doors. Paneled entry doors. R-19 and R-11 insulation.	9.60	2.84	12.44
5. Roofing	Heavyweight three tab asphalt roof shingles on 30# felt paper. Aluminum drip edge, flashings, gutters and downspouts. Metal roofing on walkout bays. Skylights.	1.21	1.23	2.44
6. Interiors	Wall finish — one coat primer and one coat paint on 1/2″ or 5/8″ gypsum wallboard. Pine door, window and baseboard moldings, one coat paint or stain. Flooring — 57% carpet, 32% vinyl, 5% ceramic tile and 6% hardwood.	6.38	5.21	11.59
7. Specialties	Hardwood faced, particle board case kitchen cabinets and bath vanities with plastic laminate countertops. Washer, dryer, cooktop with hood, double oven, dishwasher, compactor, freezer and refrigerator. Two faced masonry fireplace.	6.20	2.25	8.45
8. Mechanical	Oil fired forced hot air heat with central air conditioning. One full bath, two 1/2 baths and a master suite with a shower and whirlpool. Double bowl kitchen sink with disposal.	3.73	2.39	6.12
9. Electrical	200 amp service, branch circuit wiring with romex cable. Exterior and interior lighting fixtures, receptacles and switches.	1.91	1.03	2.94
10. Overhead	Contractor's overhead and profit.	2.88	1.74	4.62
Total Cost per Square Foot		$44.01	$26.63	$70.64

To purchase a full set of Sepias, Bill of Materials and Detailed Costs — turn to page 267.

RM-2662

2 Story Colonial

3556 Square Feet
5 Bedrooms
3½ Baths
Schedule C

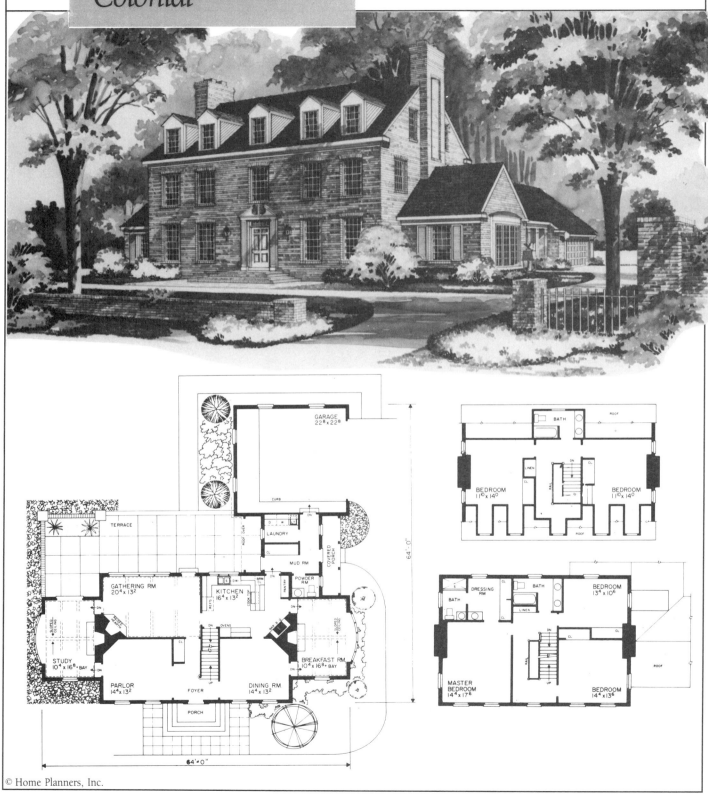

RM-2662
Cost Estimate

Cost at a Glance
Cost per Square Foot: $79.98
Total Cost: $284,408

Cost by Category

Category	Description	Cost per Square Foot of Living Area		
		Materials	Installation	Total
1. Site Work	Excavation for the basement and footings.		.56	.56
2. Foundation	Main house — 12″ wide masonry foundation wall on 20″ x 10″ reinforced concrete perimeter footings. Trench footings — 8″ and 12″ wide reinforced concrete. Slabs — 4″ thick reinforced steel trowel finished concrete over compacted gravel.	3.52	4.63	8.15
3. Framing	Exterior walls — 2 x 6 studs, 16″ on center with 1/2″ plywood sheathing. Garage — 2 x 4 studs, 16″ on center. Floors — 2 x 10 floor joists, 16″ on center with 3/4″ plywood subfloor. Roof — pre-engineered trusses and site cut rafters with 1/2″ plywood sheathing.	8.50	5.24	13.74
4. Exterior Walls	Masonry veneer and beveled cedar siding over 15# felt vapor barrier with R-19 and R-11 wall insulation. Vinyl clad double hung and casement windows and a swinging glass patio doors.	10.97	7.84	18.81
5. Roofing	Heavyweight three tab asphalt shingles over 30# felt roofing paper. Aluminum gutters, downspouts, drip edge and flashings.	1.20	1.16	2.36
6. Interiors	Walls and ceilings — 1/2″ and 5/8″ taped and finished gypsum wallboard, primed and painted with one coat latex. Pine interior trim, with one coat paint or stain. Flooring — 77% carpet, 17% vinyl, 2% hardwood and 4% ceramic tile.	7.09	6.11	13.20
7. Specialties	Hardwood faced particle board case kitchen cabinets and bathroom vanities with plastic laminate countertops. Washer, dryer, cooktop with hood, dishwasher, and refrigerator. Three masonry fireplaces and an indoor barbecue.	5.46	3.12	8.58
8. Mechanical	Oil fired forced hot air heat with central air conditioning. Two full baths, one 3/4 bath and one 1/2 bath. Stainless steel double bowl kitchen sink with disposal.	3.78	2.59	6.37
9. Electrical	200 amp service, branch circuit wiring with romex cable. Exterior and interior lighting fixtures, receptacles and switches.	1.94	1.04	2.98
10. Overhead	Contractor's overhead and profit.	2.97	2.26	5.23
	Total Cost per Square Foot	**$45.43**	**$34.55**	**$79.98**

To purchase a full set of Sepias, Bill of Materials and Detailed Costs — turn to page 267.

RM-3399

2 Story Farmhouse

3818 Square Feet
4 Bedrooms
3½ Baths
Schedule D

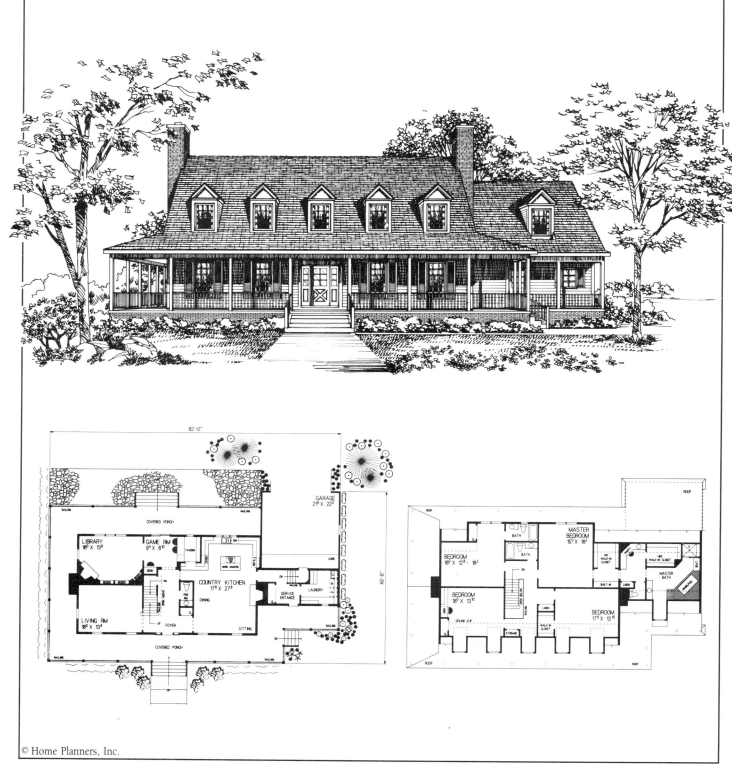

RM-3399
Cost Estimate

Cost at a Glance

Cost per Square Foot: $74.83
Total Cost: $285,700

Cost by Category

		Cost per Square Foot of Living Area		
		Materials	Installation	Total
1. Site Work	Excavation for the basement and footings.		.52	.52
2. Foundation	Main house — 12″ wide masonry foundation wall on 20″ x 10″ reinforced concrete perimeter footings. Trench footings — 8″ and 12″ wide reinforced concrete. Slabs — 4″ thick reinforced steel trowel finished concrete over compacted gravel.	3.47	4.47	7.94
3. Framing	Exterior walls — 2 x 6 studs, 16″ on center with 1/2″ plywood sheathing. Floors — 2 x 12 floor joists, 16″ on center with 3/4″ plywood subfloor. Garage — 2 x 4 studs, 16″ on center with 1/2″ plywood sheathing. Roof — site cut rafters with 1/2″ plywood sheathing.	10.03	5.47	15.50
4. Exterior Walls	Beveled cedar siding over 15# felt vapor barrier with R-19 and R-11 wall insulation. Vinyl clad double hung and casement windows and swinging glass patio doors.	8.17	4.31	12.48
5. Roofing	Heavyweight three tab asphalt shingles over 30# felt roofing paper. Aluminum gutters, downspouts, drip edge and flashings.	1.38	1.15	2.53
6. Interiors	Walls and ceilings — 1/2″ and 5/8″ taped and finished gypsum wallboard, primed and painted with one coat latex. Pine interior trim, with one coat paint or stain. Flooring — 62% carpet, 25% vinyl, 4% hardwood and 9% ceramic tile.	5.89	5.34	11.23
7. Specialties	Hardwood faced particle board case kitchen cabinets and bathroom vanities with plastic laminate countertops. Washer, dryer, cooktop with hood, dishwasher, and refrigerator. Two masonry fireplaces, covered porches.	7.34	3.22	10.56
8. Mechanical	Oil fired forced hot air heat with central air conditioning. Two full and one 1/2 baths and a master suite with a whirlpool and shower. Stainless steel double bowl kitchen sink with disposal.	3.98	2.51	6.49
9. Electrical	200 amp service, branch circuit wiring with romex cable. Exterior and interior lighting fixtures, receptacles and switches.	1.68	1.00	2.68
10. Overhead	Contractor's overhead and profit.	2.94	1.96	4.90
	Total Cost per Square Foot	$44.88	$29.95	$74.83

To purchase a full set of Sepias, Bill of Materials and Detailed Costs — turn to page 267.

RM-3414

2 Story
Southwestern

3168 Square Feet
5 Bedrooms
3½ Baths
Schedule C

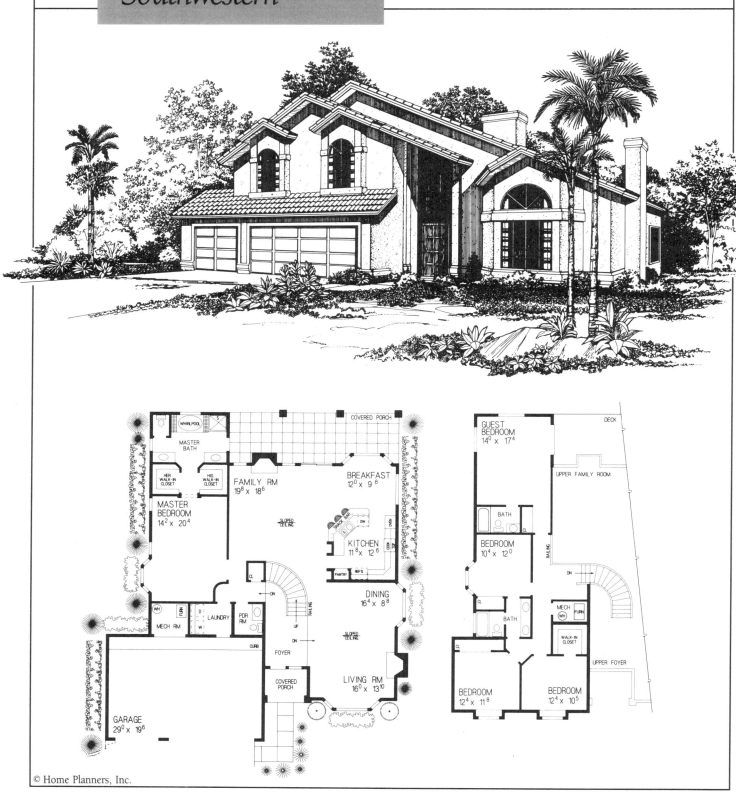

RM-3414
Cost Estimate

Cost at a Glance

Cost per Square Foot: $74.83
Total Cost: $237,061

Cost by Category

		Cost per Square Foot of Living Area		
		Materials	Installation	Total
1. Site Work	Excavation for the slab and footings.		.23	.23
2. Foundation	Main house—8″ x 24″ reinforced concrete walls on 16″ x 10″ reinforced concrete footings. Spread footings and continuous footings at interior bearing points. Slabs—4″ thick reinforced steel trowel finished concrete on 4″ compacted gravel.	1.96	2.50	4.46
3. Framing	Main house—2 x 6 studs, 16″ on center with 1/2″ plywood sheathing. Garage—2 x 4 studs, 16″ on center. Roof—pre-engineered roof trusses and 2 x 8 site cut roof rafters 5/8″ plywood sheathing.	8.63	8.75	17.38
4. Exterior Walls	Stucco on 1″ foam board siding over 30# felt vapor barrier. Vinyl clad casement windows and swinging glass doors. Paneled front entry door and flush mechanical room doors. R-19 and R-11 insulation.	10.34	2.08	12.42
5. Roofing	Concrete mission barrel type tile roofing over 30# felt roofing paper. Aluminum drip edge, flashings, gutters and downspouts.	5.61	1.67	7.28
6. Interiors	Wall finish—1/4″ (2 layers), 1/2″ or 5/8″ gypsum wallboard with one coat primer and one coat finish paint. Pine door, window and baseboard moldings. Flooring—82% carpet, 12% vinyl, 3% hardwood and 3% ceramic tile.	6.62	5.45	12.07
7. Specialties	Hardwood faced, particle board case kitchen cabinets and bath vanities with plastic laminate countertops. Washer, dryer, cooktop with hood, double ovens, microwave oven, dishwasher and refrigerator. Two pre-fabricated fireplaces, a balcony and a covered patio.	4.85	1.20	6.05
8. Mechanical	Oil fired forced hot air heat with central air conditioning. Two full baths, one 1/2 bath and a master suite with a whirlpool and shower. Double bowl kitchen sink with disposal.	4.35	2.73	7.08
9. Electrical	200 amp service, branch circuit wiring with romex cable. Exterior and interior lighting fixtures, receptacles and switches.	1.89	1.07	2.96
10. Overhead	Contractor's overhead and profit.	3.10	1.80	4.90
	Total Cost per Square Foot	**$47.35**	**$27.48**	**$74.83**

To purchase a full set of Sepias, Bill of Materials and Detailed Costs—turn to page 267.

RM-2694

2 Story
Traditional

3412 Square Feet
3 Bedrooms
2½ Baths
Schedule C

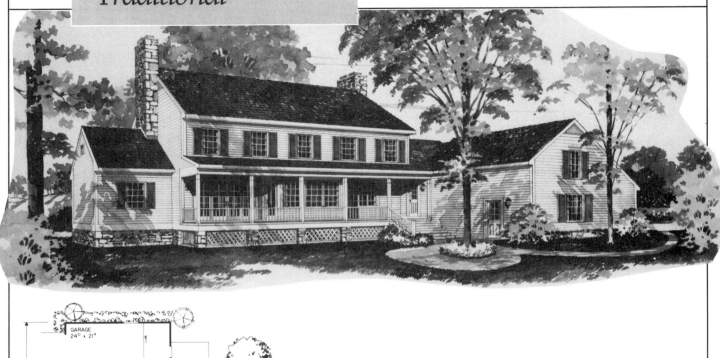

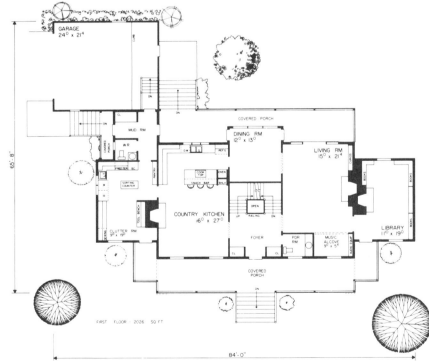

GARAGE
24⁰ x 21⁴

MUD RM

W R

PANTRY BC.

SORTING COUNTER

COOK TOP
OVEN
ONLY

SNACK BAR

CLUTTER RM
9⁰ x 9⁰

COUNTRY KITCHEN
16⁰ x 27⁰

TOOL BENCH

DINING RM
12⁰ x 13⁰

COVERED PORCH

LIVING RM
15⁰ x 21⁴

FOYER

PDR RM

MUSIC ALCOVE
9⁰ x 5⁵

LIBRARY
11⁰ x 19⁰

COVERED PORCH

FIRST FLOOR · 2026 SQ FT

65'-8"

84'-0"

SEAT   SEAT

DRESSING RM   BATH   WHIRLPOOL

BEDROOM
16⁰ x 13⁴

WALK-IN CLOSET

OPEN

RAILING

MASTER BEDROOM
16⁰ x 17⁴

LINEN

BATH

BEDROOM
12⁰ x 15⁰

SECOND FLOOR · 1386 SQ FT

RM-2694
Cost
Estimate

Cost at a Glance

Cost per Square Foot: $87.14
Total Cost: $297,321

Cost by Category

		Cost per Square Foot of Living Area		
		Materials	Installation	Total
1. Site Work	Excavation for the basement and footings.		.64	.64
2. Foundation	Main house—12″ concrete block foundation wall on 20″ x 10″ reinforced concrete footings. Garage—12″ x 42″ reinforced concrete wall. Slabs—4″ thick reinforced steel trowel finished concrete over 4″ compacted gravel.	3.48	4.54	8.02
3. Framing	Floor framing—2 x 12 floor joists with 3/4″ tongue and groove plywood subfloor. Exterior walls—2 x 6 and 2 x 4 studs, 16″ on center with 1/2″ plywood sheathing. Roof—2 x 8 and 2 x 6 site cut rafters with 1/2″ plywood sheathing.	10.48	6.62	17.10
4. Exterior Walls	Stone veneer and cedar bevel siding. Vinyl clad double hung and casement windows. Raised panel doors and fully glazed French entry doors. 15# felt vapor barrier over R-19 and R-11 insulation.	11.83	6.17	18.00
5. Roofing	Heavyweight three tab asphalt shingles over 30# felt. Aluminum drip edge, flashings, gutters and downspouts.	1.63	1.34	2.97
6. Interiors	1/2″ and 5/8″ taped and finished gypsum wallboard, primed one coat and painted one coat. 3/4″ paneling in country kitchen. Built in pine shelving. Pine door, window and baseboard moldings, painted or stained one coat. Flooring—68% carpet, 27% vinyl, 3% ceramic tile, 2% hardwood.	7.04	6.03	13.07
7. Specialties	Hardwood faced, particle board case kitchen cabinets and bath vanities, with plastic laminate countertops. Washer, dryer, cooktop with hood, double ovens, dishwasher, refrigerator and freezer. Three masonry fireplaces. Covered porches.	7.51	3.99	11.50
8. Mechanical	Oil fired forced hot air heat with central air conditioning. Two half baths, one full bath and a master bath with whirlpool, shower and two vanities. Stainless steel kitchen sink with disposal.	4.17	2.67	6.84
9. Electrical	200 amp service, branch circuit wiring with romex cable. Exterior and interior lighting fixtures, receptacles and switches.	2.10	1.20	3.30
10. Overhead	Contractor's overhead and profit.	3.38	2.32	5.70
Total Cost per Square Foot		$51.62	$35.52	$87.14

To purchase a full set of Sepias, Bill of Materials and Detailed Costs—turn to page 267.

RM-3334

2 Story Traditional

3024 Square Feet
4 Bedrooms
2½ Baths
Schedule C

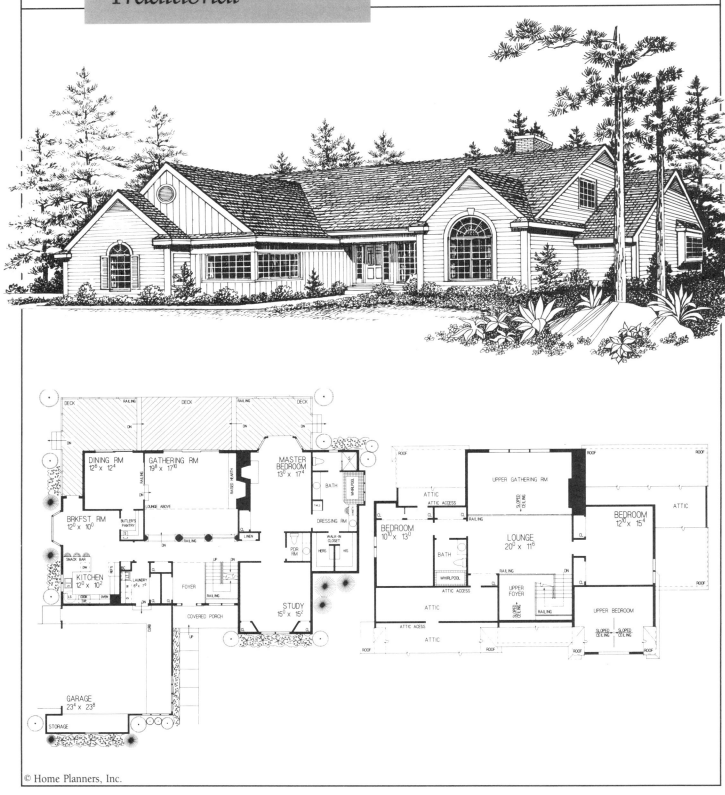

RM-3334
Cost Estimate

Cost at a Glance

Cost per Square Foot: $84.55
Total Cost: $255,679

Cost by Category

		Cost per Square Foot of Living Area		
		Materials	Installation	Total
1. Site Work	Excavation for the basement and footings.		.76	.76
2. Foundation	Main house — 10″ wide reinforced concrete foundation wall on 20″ x 10″ reinforced concrete perimeter footings. Trench footings — 8″ and 10″ wide reinforced concrete. Slabs — 4″ thick steel trowel finished reinforced concrete over compacted gravel.	6.41	8.70	15.11
3. Framing	Exterior walls — 2 x 4 studs, 16″ on center with 1/2″ plywood sheathing. Floor — 2 x 10 joists, 16″ on center with 3/4″ plywood subfloor. Roof — site cut rafters with 1/2″ plywood sheathing.	9.75	5.68	15.43
4. Exterior Walls	Beveled cedar and Texture 1-11 siding over 15# felt vapor barrier with R-19 and R-11 insulation. Vinyl clad fixed, casement and double hung windows and sliding glass patio doors.	10.64	3.72	14.36
5. Roofing	Heavyweight three tab asphalt shingles over 30# felt roofing paper. Aluminum gutters, downspouts, drip edge and flashings.	1.41	1.14	2.55
6. Interiors	Walls and ceilings — 1/2″ and 5/8″ taped and finished gypsum wallboard, primed and painted with one coat latex. Pine interior trim. Flooring — 68% carpet, 16% vinyl, 4% hardwood and 12% ceramic tile.	8.07	6.28	14.35
7. Specialties	Hardwood faced particle board case kitchen cabinets and bathroom vanities with plastic laminate countertops. Washer, dryer, cooktop with hood, double wall ovens, dishwasher and refrigerator. Two masonry fireplaces.	4.66	2.08	6.74
8. Mechanical	Oil fired forced hot air heat with central air conditioning. One full bath, one 1/2 bath and a master suite with a whirlpool and shower. Stainless steel kitchen sink with disposal.	4.24	2.64	6.88
9. Electrical	200 amp service, branch circuit wiring with romex cable. Exterior and interior lighting fixtures, receptacles and switches.	1.82	1.02	2.84
10. Overhead	Contractor's overhead and profit.	3.29	2.24	5.53
Total Cost per Square Foot		$50.29	$34.26	$84.55

To purchase a full set of Sepias, Bill of Materials and Detailed Costs — turn to page 267.

RM-2958

2 Story Tudor

3133 Square Feet
4 Bedrooms
2½ Baths
Schedule C

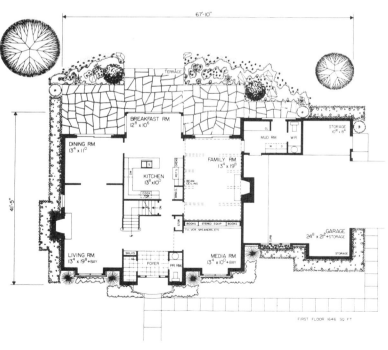

RM-2958
Cost Estimate

Cost by Category

Category	Description	Materials	Installation	Total
			Cost per Square Foot of Living Area	
1. Site Work	Excavation for the basement and footings.		.62	.62
2. Foundation	Main house — 10″ thick reinforced concrete walls on 20″ x 10″ reinforced concrete footings. Garage and porches — 10″ thick x 48″ and 54″ high reinforced concrete trench walls. Slabs — 4″ thick reinforced steel trowel finished concrete on 4″ compacted gravel.	3.33	4.30	7.63
3. Framing	Main house — 2 x 4 studs, 16″ on center with 1/2″ plywood sheathing. Garage — 2 x 4 studs, 16″ on center. Floor — 2 x 10 joists, 16″ on center with 3/4″ tongue and groove plywood subfloor. Roof — pre-engineered trusses with site cut 2 x 6 and 2 x 8 rafters, 1/2″ plywood sheathing.	7.10	4.30	11.40
4. Exterior Walls	Bevel cedar siding, stone and brick veneer and stucco over 15# felt vapor barrier. Vinyl clad double hung, casement and fixed windows and sliding glass doors. Paneled entry doors. R-19 and R-11 insulation.	10.01	6.15	16.16
5. Roofing	Heavyweight three tab asphalt roof shingles on 30# felt paper. Aluminum drip edge, flashings, gutters and downspouts.	1.24	1.01	2.25
6. Interiors	Wall finish — 1/2″ or 5/8″ gypsum wallboard. Pine door, window and baseboard moldings. Flooring — 73% carpet, 12% vinyl, 10% ceramic tile and 5% hardwood.	6.27	5.69	11.96
7. Specialties	Hardwood faced, particle board case kitchen cabinets and bath vanities with plastic laminate countertops. Washer, dryer, cooktop with hood, double ovens, dishwasher and refrigerator. Two masonry fireplaces and a terrace.	5.36	2.72	8.08
8. Mechanical	Oil fired forced hot air heat with central air conditioning. One full bath, two 1/2 baths and a master suite with a whirlpool and shower. Double bowl kitchen sink with disposal.	4.25	2.68	6.93
9. Electrical	200 amp service, branch circuit wiring with romex cable. Exterior and interior lighting fixtures, receptacles and switches.	2.13	1.17	3.30
10. Overhead	Contractor's overhead and profit.	2.78	2.00	4.78
Total Cost per Square Foot		$42.47	$30.64	$73.11

To purchase a full set of Sepias, Bill of Materials and Detailed Costs — turn to page 267.

Costs shown in Builders' Costs for 100 Best-Selling Home Plans *are based on National Averages for materials and installation. To adjust these costs to a specific location, simply multiply the base cost by the factor for that city. The data is arranged alphabetically by state and postal zip code numbers. For a city not listed, use the factor for a nearby city with similar economic characteristics.*

State/Zip	City	Factor
Alabama		
350-352	Birmingham	.82
354	Tuscaloosa	.81
355	Jasper	.76
356	Decatur	.83
357-358	Huntsville	.82
359	Gadsden	.81
360-361	Montgomery	.82
362	Anniston	.72
363	Dothan	.82
364	Evergreen	.82
365-366	Mobile	.83
367	Selma	.81
368	Phenix City	.81
369	Butler	.82
Alaska		
995-996	Anchorage	1.30
997	Fairbanks	1.28
998	Juneau	1.29
999	Ketchikan	1.34
Arizona		
850,853	Phoenix	.94
852	Mesa/Tempe	.89
855	Globe	.94
856-857	Tucson	.93
859	Show Low	.95
860	Flagstaff	.97
863	Prescott	.95
864	Kingman	.94
865	Chambers	.94
Arkansas		
716	Pine Bluff	.80
717	Camden	.73
718	Texarkana	.78
719	Hot Springs	.73

State/Zip	City	Factor
720-722	Little Rock	.81
723	West Memphis	.83
724	Jonesboro	.83
725	Batesville	.80
726	Harrison	.80
727	Fayetteville	.72
728	Russellville	.82
729	Fort Smith	.82
749	Poteau	.83
California		
900-902	Los Angeles	1.15
903-905	Inglewood	1.11
906-908	Long Beach	1.12
910-912	Pasadena	1.10
913-916	Van Nuys	1.12
917-918	Alhambra	1.12
919-921	San Diego	1.14
922	Palm Springs	1.15
923-924	San Bernardino	1.13
925	Riverside	1.17
926-927	Santa Ana	1.14
928	Anaheim	1.16
930	Oxnard	1.19
931	Santa Barbara	1.16
932-933	Bakersfield	1.15
934	San Luis Obispo	1.25
935	Mojave	1.12
936-938	Fresno	1.14
939	Salinas	1.13
940-941	San Francisco	1.24
942,956-958	Sacramento	1.12
943	Palo Alto	1.15
944	San Mateo	1.16
945	Vallejo	1.13
946	Oakland	1.17
947	Berkeley	1.29

State/Zip	City	Factor
948	Richmond	1.15
949	San Rafael	1.26
950	Santa Cruz	1.17
951	San Jose	1.23
952	Stockton	1.15
953	Modesto	1.14
954	Santa Rosa	1.18
955	Eureka	1.13
959	Marysville	1.12
960	Redding	1.12
961	Susanville	1.13
Colorado		
800-802	Denver	.97
803	Boulder	.89
804	Golden	.90
805	Fort Collins	.94
806	Greeley	.90
807	Fort Morgan	.93
808-809	Colorado Springs	.93
810	Pueblo	.93
811	Alamosa	.90
812	Salida	.90
813	Durango	.90
814	Montrose	.88
815	Grand Junction	.92
816	Glenwood Springs	.92
Connecticut		
060	New Britain	1.09
061	Hartford	1.10
062	Willimantic	1.09
063	New London	1.11
064	Meriden	1.09
065	New Haven	1.09
066	Bridgeport	1.07
067	Waterbury	1.11
068	Norwalk	1.06

State/Zip	City	Factor
Connecticut (cont'd)		
069	Stamford	1.08
D.C.		
200-205	Washington	.95
Delaware		
197	Newark	1.01
198	Wilmington	1.00
199	Dover	1.01
Florida		
320,322	Jacksonville	.87
321	Daytona Beach	.91
323	Tallahassee	.80
324	Panama City	.73
325	Pensacola	.89
326	Gainesville	.88
327-328,347	Orlando	.91
329	Melbourne	.92
330-332,340	Miami	.87
333	Fort Lauderdale	.87
334,349	West Palm Beach	.90
335-336,346	Tampa	.85
337	St. Petersburg	.72
338	Lakeland	.84
339	Fort Myers	.85
342	Sarasota	.84
Georgia		
300-303,399	Atlanta	.82
304	Statesboro	.67
305	Gainesville	.62
306	Athens	.73
307	Dalton	.68
308-309	Augusta	.78
310-312	Macon	.82
313-314	Savannah	.82

State/Zip	City	Factor
315	Waycross	.76
316	Valdosta	.77
317	Albany	.78
318-319	Columbus	.78
Hawaii		
967	Hilo	1.27
968	Honolulu	1.26
States & Poss.		
969	Guam	.86
Idaho		
832	Pocatello	.93
833	Twin Falls	.83
834	Idaho Falls	.86
835	Lewiston	1.11
836-837	Boise	.93
838	Coeur d'Alene	1.00
Illinois		
600-603	North Suburban	1.05
604	Joliet	1.06
605	South Suburban	1.02
606	Chicago	1.10
609	Kankakee	.95
610-611	Rockford	1.00
612	Rock Island	.99
613	La Salle	1.03
614	Galesburg	1.01
615-616	Peoria	1.03
617	Bloomington	.98
618-619	Champaign	1.01
620-622	East St. Louis	.96
623	Quincy	.93
624	Effingham	.95
625	Decatur	.97
626-627	Springfield	.96

State/Zip	City	Factor
628	Centralia	.94
629	Carbondale	.91
Indiana		
424	Henderson	.93
460	Anderson	.93
461-462	Indianapolis	.97
463-464	Gary	.98
465-466	South Bend	.92
467-468	Fort Wayne	.90
469	Kokomo	.89
470	Lawrenceburg	.89
471	New Albany	.89
472	Columbus	.90
473	Muncie	.92
474	Bloomington	.92
475	Washington	.90
476-477	Evansville	.92
478	Terre Haute	.93
479	Lafayette	.88
Iowa		
500-503,509	Des Moines	.93
504	Mason City	.89
505	Fort Dodge	.83
506-507	Waterloo	.89
508	Creston	.93
510-511	Sioux City	.89
512	Sibley	.83
513	Spencer	.84
514	Carroll	.89
515	Council Bluffs	.93
516	Shenandoah	.77
520	Dubuque	.94
521	Decorah	.93
522-524	Cedar Rapids	.98
525	Ottumwa	.95

State/Zip	City	Factor
Iowa (cont'd)		
526	Burlington	.84
527-528	Davenport	.91
Kansas		
660-662	Kansas City	.95
664-666	Topeka	.87
667	Fort Scott	.88
668	Emporia	.79
669	Belleville	.92
670-672	Wichita	.88
673	Independence	.82
674	Salina	.87
675	Hutchinson	.81
676	Hays	.90
677	Colby	.90
678	Dodge City	.89
679	Liberal	.82
Kentucky		
400-402	Louisville	.93
403-405	Lexington	.90
406	Frankfort	.96
407-409	Corbin	.82
410	Covington	.96
411-412	Ashland	.94
413-414	Campton	.80
415-416	Pikeville	.85
417-418	Hazard	.80
420	Paducah	.95
421-422	Bowling Green	.93
423	Owensboro	.91
425-426	Somerset	.79
427	Elizabethtown	.92
Lousiana		
700-701	New Orleans	.88
703	Thibodaux	.87

State/Zip	City	Factor
704	Hammond	.83
705	Lafayette	.86
706	Lake Charles	.87
707-708	Baton Rouge	.86
710-711	Shreveport	.82
712	Monroe	.80
713-714	Alexandria	.80
Maine		
039	Kittery	.82
040-041	Portland	.89
042	Lewiston	.91
043	Augusta	.82
044	Bangor	.93
045	Bath	.82
046	Machias	.73
047	Houlton	.84
048	Rockland	.87
049	Waterville	.83
Maryland		
206	Waldorf	.88
207-208	College Park	.91
209	Silver Spring	.90
210-212	Baltimore	.91
214	Annapolis	.90
215	Cumberland	.87
216	Easton	.67
217	Hagerstown	.88
218	Salisbury	.78
219	Elkton	.80
Massachusetts		
010-011	Springfield	1.09
012	Pittsfield	1.05
013	Greenfield	1.06
014	Fitchburg	1.12

State/Zip	City	Factor
015-016	Worcester	1.13
017	Framingham	1.11
018	Lowell	1.13
019	Lawrence	1.14
020-022	Boston	1.20
023-024	Brockton	1.11
025	Buzzards Bay	1.07
026	Hyannis	1.09
027	New Bedford	1.10
Michigan		
480,483	Royal Oak	1.02
481	Ann Arbor	1.02
482	Detroit	1.06
484-485	Flint	.98
486	Saginaw	.96
487	Bay City	.95
488-489	Lansing	.97
490	Battle Creek	.98
491	Kalamazoo	.98
492	Jackson	.93
493,495	Grand Rapids	.90
494	Muskegan	.96
496	Traverse City	.92
497	Gaylord	.93
498-499	Iron Mountain	.96
Minnesota		
540	New Richmond	.93
550-551	Saint Paul	1.03
553-554	Minneapolis	1.09
556-558	Duluth	.94
559	Rochester	1.01
560	Mankato	.93
561	Windom	.79
562	Willmar	.82
563	St. Cloud	1.02
564	Brainerd	.99

State/Zip	City	Factor
Minnesota (cont'd)		
565	Detroit Lakes	.85
566	Bemidji	.84
567	Thief River Falls	.81
Mississippi		
386	Clarksdale	.71
387	Greenville	.82
388	Tupelo	.72
389	Greenwood	.73
390-392	Jackson	.83
393	Meridian	.77
394	Laurel	.73
395	Biloxi	.84
396	Mc Comb	.70
397	Columbus	.72
Missouri		
630-631	St. Louis	.98
633	Bowling Green	.92
634	Hannibal	1.00
635	Kirksville	.85
636	Flat River	.92
637	Cape Girardeau	.91
638	Sikeston	.85
639	Poplar Bluff	.85
640-641	Kansas City	.97
644-645	St. Joseph	.86
646	Chillicothe	.80
647	Harrisonville	.92
648	Joplin	.85
650-651	Jefferson City	.95
652	Columbia	.94
653	Sedalia	.92
654-655	Rolla	.89
656-658	Springfield	.83

State/Zip	City	Factor
Montana		
590-591	Billings	.99
592	Wolf Point	.94
593	Miles City	.95
594	Great Falls	.98
595	Havre	.92
596	Helena	.95
597	Butte	.93
598	Missoula	.96
599	Kalispell	.96
Nebraska		
680-681	Omaha	.89
683-685	Lincoln	.86
686	Columbus	.76
687	Norfolk	.86
688	Grand Island	.85
689	Hastings	.85
690	Mc Cook	.76
691	North Platte	.85
692	Valentine	.80
693	Alliance	.76
Nevada		
889-891	Las Vegas	1.03
893	Ely	.95
894-895	Reno	.94
897	Carson City	.96
898	Elko	.93
New Hampshire		
030	Nashua	.95
031	Manchester	.95
032-033	Concord	.93
034	Keene	.84
035	Littleton	.86
036	Charleston	.82

State/Zip	City	Factor
037	Claremont	.81
038	Portsmouth	.94
New Jersey		
070-071	Newark	1.08
072	Elizabeth	1.07
073	Jersey City	1.07
074-075	Paterson	1.07
076	Hackensack	1.06
077	Long Branch	1.06
078	Dover	1.07
079	Summit	1.06
080,083	Vineland	1.04
081	Camden	1.04
082,084	Atlantic City	1.06
085-086	Trenton	1.07
087	Point Pleasant	1.07
088-089	New Brunswick	1.08
New Mexico		
870-872	Albuquerque	.88
873	Gallup	.90
874	Farmington	.89
875	Santa Fe	.88
877	Las Vegas	.88
878	Socorro	.90
879	Truth/ Consequences	.89
880	Las Cruces	.84
881	Clovis	.90
882	Roswell	.91
883	Carrizozo	.92
884	Tucumcari	.91
New York		
100-102	New York	1.34
103	Staten Island	1.26

State/Zip	City	Factor
New York (cont'd)		
104	Bronx	1.25
105	Mount Vernon	1.24
106	White Plains	1.20
107	Yonkers	1.23
108	New Rochelle	1.24
109	Suffern	1.10
110	Queens	1.25
111	Long Island City	1.26
112	Brooklyn	1.26
113	Flushing	1.26
114	Jamaica	1.25
115,117,118	Hicksville	1.16
116	Far Rockaway	1.26
119	Riverhead	1.16
120-122	Albany	.98
123	Schenectady	.99
124	Kingston	1.12
125-126	Poughkeepsie	1.12
127	Monticello	1.11
128	Glens Falls	.97
130-132	Syracuse	1.00
133-135	Utica	.90
136	Watertown	.93
137-139	Binghamton	.94
140-142	Buffalo	1.08
143	Niagara Falls	1.03
144-146	Rochester	1.00
147	Jamestown	.94
148-149	Elmira	.95
North Carolina		
270,272-274	Greensboro	.77
271	Winston-Salem	.77
275-276	Raleigh	.78
277	Durham	.77
278	Rocky Mount	.65

State/Zip	City	Factor
279	Elizabeth City	.67
280	Gastonia	.77
281-282	Charlotte	.77
283	Fayetteville	.77
284	Wilmington	.75
285	Kinston	.66
286	Hickory	.64
287-288	Asheville	.75
289	Murphy	.64
North Dakota		
580-581	Fargo	.79
582	Grand Forks	.80
583	Devils Lake	.80
584	Jamestown	.80
585	Bismarck	.81
586	Dickinson	.81
587	Minot	.80
588	Williston	.80
Ohio		
430-432	Columbus	.95
433	Marion	.89
434-436	Toledo	.97
437-438	Zanesville	.88
439	Steubenville	.93
440	Lorain	.98
441	Cleveland	1.08
442-443	Akron	.99
444-445	Youngstown	.98
446-447	Canton	.95
448-449	Mansfield	.91
450	Hamilton	.96
451-452	Cincinnati	.98
453-454	Dayton	.91
455	Springfield	.89
456	Chillicothe	.96

State/Zip	City	Factor
457	Athens	.90
458	Lima	.89
Oklahoma		
730-731	Oklahoma City	.82
734	Ardmore	.82
735	Lawton	.82
736	Clinton	.79
737	Enid	.81
738	Woodward	.81
739	Guymon	.72
740-741	Tulsa	.87
743	Miami	.84
744	Muskogee	.78
745	Mc Alester	.77
746	Ponca City	.80
747	Durant	.78
748	Shawnee	.79
Oregon		
970-972	Portland	1.07
973	Salem	1.05
974	Eugene	1.05
975	Medford	1.05
976	Klamath Falls	1.05
977	Bend	1.05
978	Pendleton	1.02
979	Vale	.98
Pennsylvania		
150-152	Pittsburgh	1.03
153	Washington	.98
154	Uniontown	.98
155	Bedford	1.00
156	Greensburg	.98
157	Indiana	1.02
158	Dubois	1.01
159	Johnstown	1.04

State/Zip	City	Factor
Pennsylvania (cont'd)		
160	Butler	.97
161	New Castle	.99
162	Kittanning	1.00
163	Oil City	.89
164-165	Erie	.96
166	Altoona	1.05
167	Bradford	.97
168	State College	.97
169	Wellsboro	.91
170-171	Harrisburg	.97
172	Chambersburg	.94
173-174	York	.95
175-176	Lancaster	.95
177	Williamsport	.92
178	Sunbury	.93
179	Pottsville	.93
180	Lehigh Valley	.99
181	Allentown	1.03
182	Hazleton	.95
183	Stroudsburg	.99
184-185	Scranton	.96
186-187	Wilkes-Barre	.94
188	Montrose	.92
189	Doylestown	.92
190-191	Philadelphia	1.14
193	Westchester	1.03
194	Norristown	1.05
195-196	Reading	.97
Rhode Island		
028	Newport	1.03
029	Providence	1.04
South Carolina		
290-292	Columbia	.75
293	Spartanburg	.74
294	Charleston	.76

State/Zip	City	Factor
295	Florence	.73
296	Greenville	.74
297	Rock Hill	.66
298	Aiken	.66
299	Beaufort	.70
South Dakota		
570-571	Sioux Falls	.86
572	Watertown	.85
573	Mitchell	.85
574	Aberdeen	.86
575	Pierre	.86
576	Mobridge	.86
577	Rapid City	.86
Tennessee		
370-372	Nashville	.82
373-374	Chattanooga	.85
375,380-381	Memphis	.86
376	Johnson City	.81
377-379	Knoxville	.81
382	Mc Kenzie	.71
383	Jackson	.69
384	Columbia	.77
385	Cookeville	.70
Texas		
750	Mc Kinney	.92
751	Waxahackie	.83
752-753	Dallas	.90
754	Greenville	.81
755	Texarkana	.91
756	Longview	.86
757	Tyler	.91
758	Palestine	.76
759	Lufkin	.81
760-761	Fort Worth	.85

State/Zip	City	Factor
762	Denton	.89
763	Wichita Falls	.82
764	Eastland	.78
765	Temple	.81
766-767	Waco	.83
768	Brownwood	.76
769	San Angelo	.82
770-772	Houston	.90
773	Huntsville	.84
774	Wharton	.79
775	Galveston	.88
776-777	Beaumont	.87
778	Bryan	.84
779	Victoria	.85
780	Laredo	.80
781-782	San Antonio	.83
783-784	Corpus Christi	.83
785	Mc Allen	.83
786-787	Austin	.81
788	Del Rio	.72
789	Giddings	.78
790-791	Amarillo	.83
792	Childress	.79
793-794	Lubbock	.82
795-796	Abilene	.80
797	Midland	.83
798-799,885	El Paso	.80
Utah		
840-841	Salt Lake City	.88
842,844	Ogden	.88
843	Logan	.90
845	Price	.83
846-847	Provo	.89
Vermont		
050	White River Jct.	.76
051	Bellows Falls	.77

State/Zip	City	Factor
Vermont (cont'd)		
052	Bennington	.73
053	Brattleboro	.77
054	Burlington	.85
056	Montpelier	.84
057	Rutland	.87
058	St. Johnsbury	.78
059	Guildhall	.77
129	Plattsburgh	.94
Virginia		
220-221	Fairfax	.88
222	Arlington	.89
223	Alexandria	.90
224-225	Fredericksburg	.82
226	Winchester	.81
227	Culpeper	.79
228	Harrisonburg	.77
229	Charlottesville	.84
230-232	Richmond	.85
233-235	Norfolk	.82
236	Newport News	.83
237	Portsmouth	.80
238	Petersburg	.85
239	Farmville	.76
240-241	Roanoke	.80
242	Bristol	.81
243	Pulaski	.73
244	Staunton	.75
245	Lynchburg	.82
246	Grundy	.72
Washington		
980-981,987	Seattle	1.02
982	Everett	.96
983-984	Tacoma	1.07

State/Zip	City	Factor
985	Olympia	1.07
986	Vancouver	1.09
988	Wenatchee	.98
989	Yakima	1.04
990-992	Spokane	1.01
993	Richland	1.01
994	Clarkston	1.01
West Virginia		
247-248	Bluefield	.82
249	Lewisburg	.84
250-253	Charleston	.91
254	Martinsburg	.77
255-257	Huntington	.90
258-259	Beckley	.82
260	Wheeling	.90
261	Parkersburg	.88
262	Buckhannon	.93
263-264	Clarksburg	.93
265	Morgantown	.94
266	Gassaway	.91
267	Romney	.87
268	Petersburg	.93
Wisconsin		
530,532	Milwaukee	.99
531	Kenosha	.94
534	Racine	.98
535	Beloit	.91
537	Madison	.93
538	Lancaster	.90
539	Portage	.87
541-543	Green Bay	.96
544	Wausau	.90
545	Rhinelander	.94
546	La Crosse	.91
547	Eau Claire	.96

State/Zip	City	Factor
548	Superior	.98
549	Oshkosh	.92
Wyoming		
820	Cheyenne	.87
821	Yellow. Nat'l Park	.81
822	Wheatland	.84
823	Rawlins	.85
824	Worland	.82
825	Riverton	.83
826	Casper	.87
827	Newcastle	.83
828	Sheridan	.86
829-831	Rock Springs	.85

Canadian Factors
(reflect Canadian currency)

Alberta

	City	Factor
	Calgary	1.05
	Edmonton	1.05

British Columbia

	City	Factor
	Vancouver	1.09
	Victoria	1.06

Manitoba

	City	Factor
	Winnipeg	1.03

New Brunswick

	City	Factor
	Moncton	.98
	Saint John	1.01

Newfoundland

	City	Factor
	St. John's	.99

State/Zip	City	Factor
Nova Scotia		
	Halifax	1.01
Ontario		
	Hamilton	1.17
	Kitchener	1.08
	London	1.13
	Oshawa	1.13
	Ottawa	1.14
	St. Catherines	1.09
	Sudbury	1.08
	Thunder Bay	1.10
	Toronto	1.15
	Windsor	1.10

State/Zip	City	Factor
Prince Edward Island		
	Charlottetown	.96
Quebec		
	Chicoutimi	1.06
	Montreal	1.12
	Quebec	1.14
Saskatchewan		
	Regina	.94
	Saskatoon	.94

Ordering Information

The Basic Builder's Package

The Builder's Spec Plan Package includes three vital components — the Sepia Set, the Materials List, and the Detailed Cost Estimate. Together, these components form the basis for your building plan.

The Builder's Spec Plan Package contains everything you need for building homes that will sell. Hundreds of hours of painstaking effort have gone into the development of these plans and each has been quality-checked by experienced professionals to ensure accuracy and buildability.

The package includes:

- Sepia Sets
- Frontal Sheet
- Foundation Plan
- Detailed Floor Plan
- House Cross Sections
- Interior Elevations
- Exterior Elevations

Once you have made a decision about the package you would like, call our toll-free hotline to order: 1-800-521-6797, or fill in the order form on page 267.

Plans & Elevations

Sepia Sets

The sepia sets contain a variety of interrelated detail sheets which show floor plans, interior and exterior elevations, dimensions, cross sections, diagrams and notations. All are printed on reproducible sepia paper that can be easily altered for modifying or customizing the plans. Among the sheets included may be:

Frontal Sheet This artist's sketch of the exterior of the house gives you an idea of how the house will look when built and landscaped. Large ink-line floor plans show all levels of the house and provide an overview of the home's livability, as well as a handy reference for deciding on furniture placement.

Foundation Plan This sheet shows the foundation layout including support walls, excavated and unexcavated areas, if any, and foundation notes. If slab construction rather than basement, the plan shows footings and details for a monolithic slab. This page, or another in the set, may include a sample plot plan for locating the house on a building site.

Detailed Floor Plans These plans show the layout of each floor of the house. Rooms and interior spaces are carefully dimensioned and keys are given for cross-section details given later in the plans. The positions of electrical outlets and switches are shown.

House Cross Sections Large-scale views show sections or cut-aways of the foundation, interior walls, exterior walls, floors, stairways and roof details. Additional cross sections may show important changes in floor, ceiling or roof heights or the relationship of one level to another. Extremely valuable for construction, these sections show exactly how the various parts of the house fit together.

Interior Elevations These large-scale drawings show the design and placement of kitchen and bathroom cabinets, laundry areas, fireplaces, bookcases, and other built-ins. Little "extras," such as mantlepiece and wainscoting drawings, plus moulding sections, provide details that give a home that custom touch.

Exterior Elevations These drawings show the front, rear and sides of the house and give necessary notes on exterior materials and finishes. Particular attention is given to cornice detail, brick and stone accents or other finish items that make the home unique.

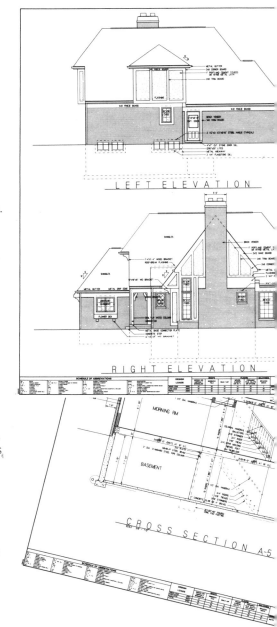

INTERIOR ELEVATIONS

FRONT ELEVATION

REAR ELEVATION

CROSS SECTION B-5

FIRST FLOOR PLAN

FOUNDATION PLAN

HOME PLANNERS, INC. 2491 3/5

HOME PLANNERS, INC. 2491 5/5

HOME PLANNERS, INC. 2491 2/5

HOME PLANNERS, INC. 2491 1/5

Materials List

The Materials List

Our customized materials takeoff is invaluable in planning and estimating the cost of building the home. The comprehensive list outlines the quantity, type and size of materials needed to build the house (with the exception of some items in the mechanical and electrical systems).

This handy list, along with the Detailed Cost Estimate, helps you cost out materials and serves as a ready reference sheet when you're compiling bids and working with subcontractors. It also helps you coordinate the substitution of items you may need to meet local codes.

(**Note:** *Because of differing local codes, building methods, and availability of materials, our Materials Lists do not include some mechanical and electrical materials. To obtain necessary takeoffs and recommendations, consult heating, plumbing and electrical contractors.*)

Included are:

- Framing lumber
- Roofing and sheet metal
- Windows and doors
- Exterior sheathing material and trim
- Masonry, veneer, and fireplace materials
- Tile and flooring materials
- Kitchen and bath cabinetry
- Interior drywall and trim
- Rough and finish hardware
- Many more items

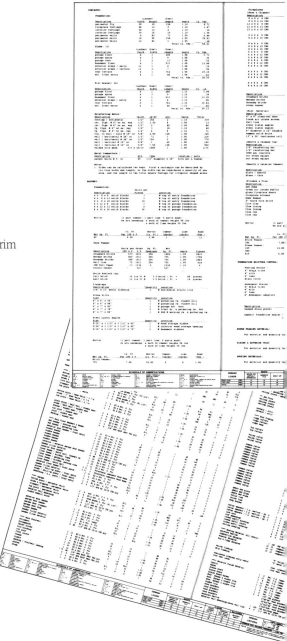

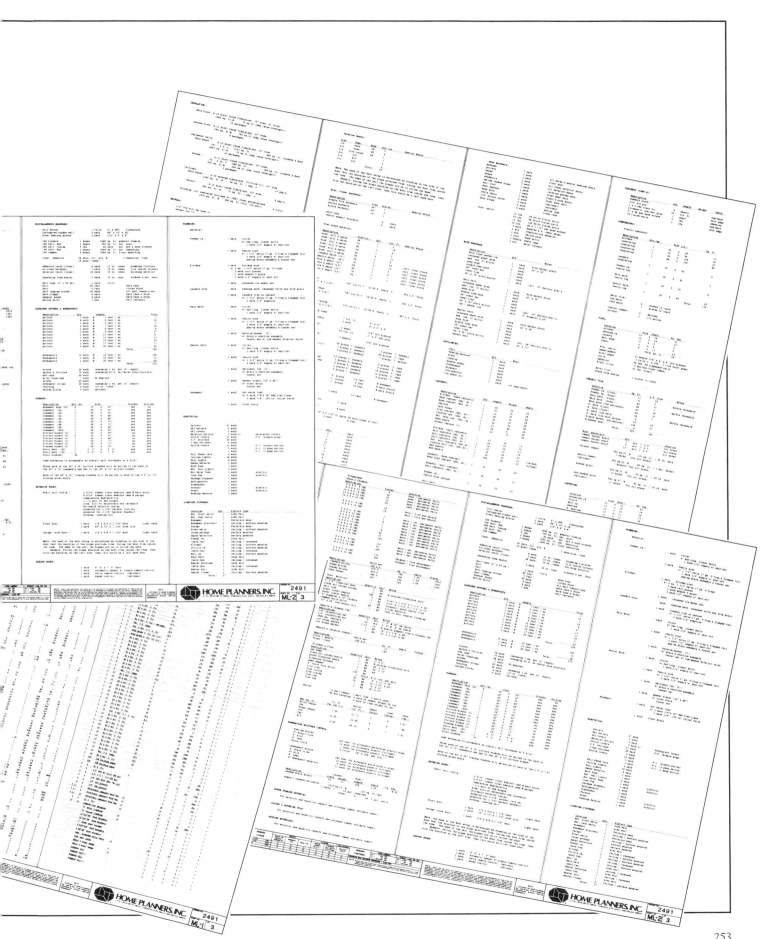

Detailed Cost Estimates

The Detailed Cost Estimate provides localized, zip-code based information on the costs of materials and labor — all custom-calculated for a specific plan in your specific area.

Some of the many categories included in the Detailed Cost Estimate are:

- Lumber
- Concrete
- Windows
- Interior and exterior doors
- Plumbing
- Electrical
- Painting
- Appliances
- Cabinets
- Flooring
- And much, much more!

The Detailed Cost Estimate is a companion to the Materials List and is a key part of your Builder's Spec Plan Package. Along with the Sepia Sets, this plan package provides the most complete, specific and detailed information ever compiled for the spec builder. It is unique, state of the art and, most importantly, invaluable as a reference and planning tool.

Each building category is broken down into line-item detail that matches the Materials List. The materials and installation costs (labor + equipment) are shown for each one of over 1,000 line items. Of particular note is the Lumber Report which lists each item of lumber needed including rafters, joists, ledgers and decking. Prices for both materials and labor are given for each item and then totaled.

The Detailed Cost Estimate also provides space for writing in prices of differing grade options should you choose to vary your choices for each line item. In addition, the Detailed Cost Estimate furnishes a Summary Cost and Bid Comparison Worksheet that lists each category, gives its total estimate and allows space for bids from subcontractors so that you can accurately compare costs.

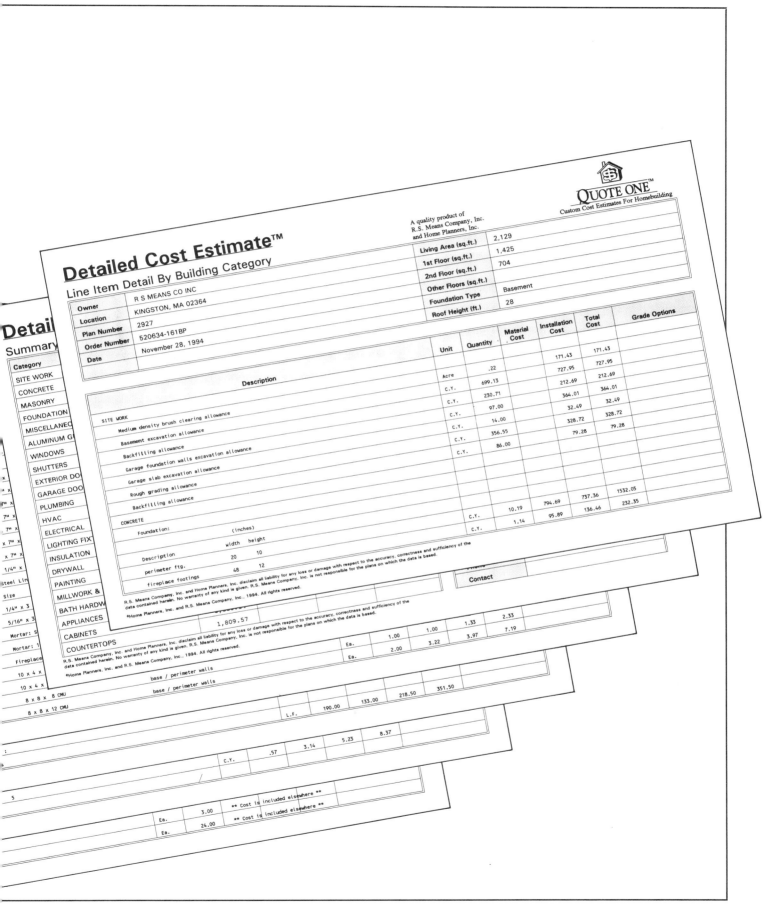

Detailed Cost Estimate™

Line Item Detail By Building Category

QUOTE ONE™
Custom Cost Estimates For Homebuilding

Living Area (sq.ft.)	2,129
1st Floor (sq.ft.)	1,425
2nd Floor (sq.ft.)	704
Other Floors (sq.ft.)	
Foundation Type	Basement
Roof Height (ft.)	28

Owner	R S MEANS CO INC
Location	KINGSTON, MA 02364
Plan Number	2927
Order Number	520634-161BP
Date	November 28, 1994

Description	Unit	Quantity	Material Cost	Installation Cost	Total Cost	Grade Options
				171.43	171.43	
	Acre	.22		727.95	727.95	
	C.Y.	699.13		212.69	212.69	
	C.Y.	230.71		364.01	364.01	
SITE WORK	C.Y.	97.00		32.49	32.49	
Medium density brush clearing allowance	C.Y.	14.00		328.72	328.72	
Basement excavation allowance	C.Y.	356.55		79.28	79.28	
Backfilling allowance	C.Y.	86.00				
Garage foundation walls excavation allowance						
Garage slab excavation allowance						
Rough grading allowance						
Backfilling allowance			794.69	737.36	1532.05	
CONCRETE	C.Y.	10.19	95.89	136.46	232.35	
Foundation: (inches)	C.Y.	1.14				

Description	width	height
perimeter ftg.	20	10
fireplace footings	48	12

Detail

Summary

Category
SITE WORK
CONCRETE
MASONRY
FOUNDATION
MISCELLANEO
ALUMINUM G
WINDOWS
SHUTTERS
EXTERIOR DO
GARAGE DOO
PLUMBING
HVAC
ELECTRICAL
LIGHTING FIX
INSULATION
DRYWALL
PAINTING
MILLWORK &
BATH HARDW
APPLIANCES
CABINETS
COUNTERTOPS

				Contact	
			1.33	2.33	
		1.00	1.00	3.97	7.19
	Ea.	2.00	3.22		

1,809.57

	base / perimeter walls					
10 x 4 x	base / perimeter walls					
8 x 8 x 8 CMU			218.50	351.50	
8 x 8 x 12 CMU	L.F.	190.00	133.00		

| | C.Y. | .57 | 3.14 | 5.23 | 8.37 |

| | Ea. | 3.00 | ** Cost is included elsewhere ** |
| | Ea. | 24.00 | ** Cost is included elsewhere ** |

Builder's Option Package

Important Extras to Do the Job Right!
Introducing invaluable planning and construction aids developed by our professionals to help you in your spec building project.

Specification Outline

This valuable 16-page document is critical to building your house correctly. The book lists 166 stages or items crucial to the building process. It provides a comprehensive review of the construction process and helps in making choices of materials. It can serve as a guide for preparing a quote and forms the basis for the construction program.

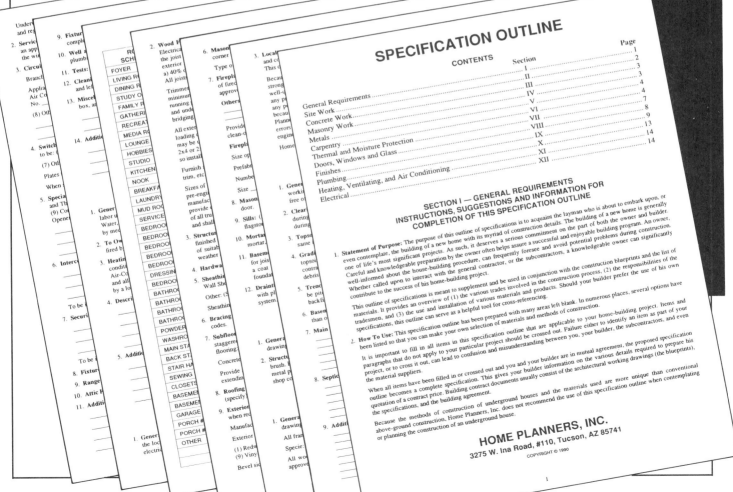

SPECIFICATION OUTLINE

SPECIFICATION OUTLINE

SECTION I — GENERAL REQUIREMENTS
INSTRUCTIONS, SUGGESTIONS AND INFORMATION FOR
COMPLETION OF THIS SPECIFICATION OUTLINE

1. **Statement of Purpose:** The purpose of this outline of specifications is to acquaint the layman who is about to embark upon, or even contemplate, the building of a new home with its myriad of construction details. The building of a new home is generally one of life's most significant projects. As such, it deserves a serious commitment on the part of both the owner and builder. Careful and knowledgeable preparation by the owner often helps assure a successful and enjoyable building program. An owner, well-informed about the house-building procedure, can frequently foresee and avoid potential problems during construction. Whether called upon to interact with the general contractor, or the subcontractors, a knowledgeable owner can significantly contribute to the success of his home-building project.

This outline of specifications is meant to supplement and be used in conjunction with the construction blueprints and the list of materials. It provides an overview of (1) the various trades involved in the construction process, (2) the responsibilities of the tradesmen, and (3) the use and installation of various materials and products. Should your builder prefer the use of his own specifications, this outline can serve as a helpful tool for cross-referencing.

2. **How To Use:** This specification outline has been prepared with many areas left blank. In numerous places, several options have been listed so that you can make your own selection of materials and methods of construction.

It is important to fill in all items in this specification outline that are applicable to your home-building project. Items and paragraphs that do not apply to your particular project should be crossed out. Failure either to identify an item as part of your project, or to cross it out, can lead to confusion and misunderstanding between you, your builder, the subcontractors, and even the material suppliers.

When all items have been filled in or crossed out and you and your builder are in mutual agreement, the proposed specification outline becomes a complete specification. This gives your builder information on the various details required to prepare his quotation of a contract price. Building contract documents usually consist of the architectural working drawings (the blueprints), the specifications, and the building agreement.

Because the methods of construction of underground houses and the materials used are more unique than conventional above-ground construction, Home Planners, Inc. does not recommend the use of this specification outline when contemplating or planning the construction of an underground house.

HOME PLANNERS, INC.
3275 W. Ina Road, #110, Tucson, AZ 85741
COPYRIGHT © 1990

1

Construction Detail Information

These remarkably useful detail sheets will enhance your understanding of technical subjects.

Plumbing The Sepia Package contains locations for all the plumbing fixtures in the house, including sinks, lavatories, tubs, showers, toilets, laundry trays and water heaters. However, if you want to know more about plumbing systems in general, these 24 x 36-inch detail sheets will prove very useful. Prepared to meet requirements of the National Plumbing Code, these six fact-filled sheets give standard, generic information on pipe schedules, fittings, sump-pump details, water-softener hookups, septic system details and much more. Color-coded sheets include a glossary of terms.

Construction The Sepia Set contains everything an experienced builder needs to construct a particular house. However, it doesn't show all the ways houses can be built, nor does it explain alternate construction methods. Designed as a helpful guide to standard building techniques, this set of drawings depicts the materials and methods used to build foundations, fireplaces, walls, floors and roofs. Where appropriate, these six sheets show acceptable alternatives.

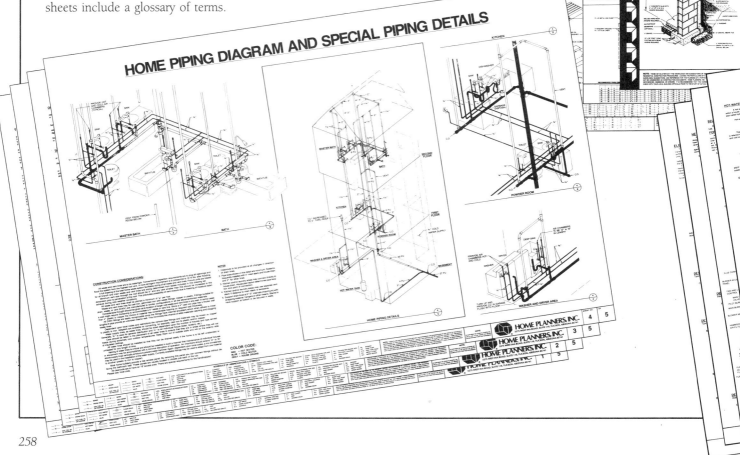

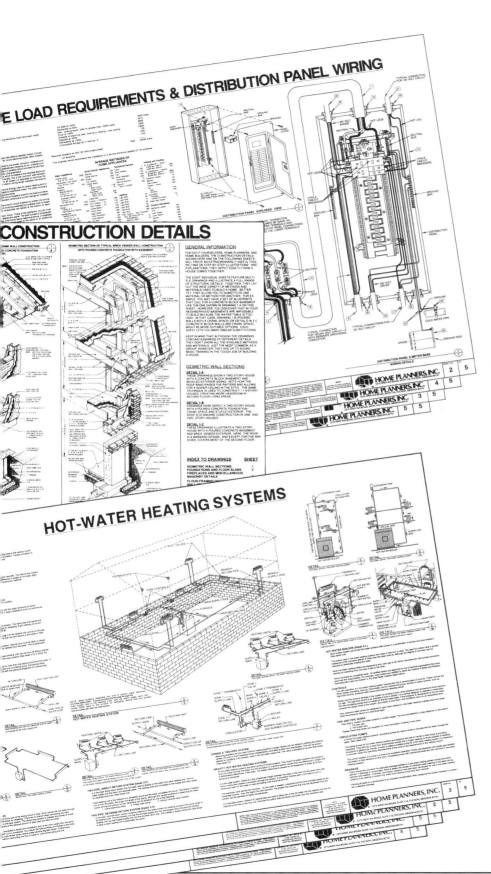

Electrical The locations for every electrical switch, plug and outlet are shown in the Sepia Set. However, these Electrical Details go further to take the mystery out of standard household electrical systems. Prepared to meet requirements of the National Electrical Code, these comprehensive 24 x 36-inch drawings come packed with helpful information, including wire sizing, switch-installation schematics, cable-routing details, appliance wattage, door-bell hookups, typical service panel circuitry and much more. Six sheets are bound together and color-coded for easy reference. A glossary of terms is also included.

Mechanical This package contains fundamental principles and useful data that will help you make informed decisions and communicate with subcontractors about heating and cooling systems. The 24 x 36-inch drawings contain instructions and samples that allow you to make simple load calculations and do preliminary sizing and costing analyses. Covered are today's most commonly used systems from heat pumps to gas-fired furnaces and air-conditioners as well as solar fuel systems. The package is packed full of illustrations and diagrams to help you visualize components and how they relate to one another.

Landscape Package

The Landscape Blueprint Package

For the homes indicated under the "Landscape" column of the Plan Index on pages 264–265 of this book, Home Planners has created a front-yard landscape plan that matches or is complementary in design to the house plan. These comprehensive blueprint packages include a Frontal Sheet, Plan View, Regionalized Plant & Materials List, a sheet on Planting and Maintaining Your Landscape, Zone Maps and Plant Size and Description Guide. These plans will help you achieve professional results, adding value and enjoyment to your property for years to come. Each set of blueprints is a full 18″ × 24″ in size with clear, complete instructions and easy-to-read type. Six of the forty front-yard Landscape Plans available to match your favorite house are shown below.

Regional Order Map

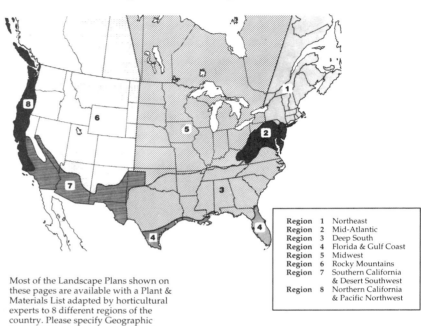

Region	1	Northeast
Region	2	Mid-Atlantic
Region	3	Deep South
Region	4	Florida & Gulf Coast
Region	5	Midwest
Region	6	Rocky Mountains
Region	7	Southern California & Desert Southwest
Region	8	Northern California & Pacific Northwest

Most of the Landscape Plans shown on these pages are available with a Plant & Materials List adapted by horticultural experts to 8 different regions of the country. Please specify Geographic Region when ordering your plan. See page 186 for prices, ordering information and regional availability.

CAPE COD COTTAGE
Landscape Plan L202

GAMBREL–ROOF COLONIAL
Landscape Plan L203

CENTER–HALL COLONIAL
Landscape Plan L204

CLASSIC NEW ENGLAND COLONIAL
Landscape Plan L205

COUNTRY–STYLE FARMHOUSE
Landscape Plan L207

TRADITIONAL SPLIT–LEVEL
Landscape Plan L228

Deck Package

The Deck Blueprint Package

Many of the homes in this book can be enhanced with a professionally designed Home Planners' Deck Plan. Those home plans indicated under the "Deck" column on the Plan Index on pages 264–265 have a matching or corresponding custom deck plan available which includes a Deck Plan Frontal Sheet, Deck Framing and Floor Plans, Deck Elevations and a Deck Materials List. A Standard Deck Details Package is also included, which provides all the how-to information necessary for building *any* deck. The plans and details are carefully prepared in an easy-to-understand format that will guide you through every stage of your deck-building project. This page contains a sampling of 12 of the 25 different Deck layouts available to match your favorite house. See pages 264–267 for prices and ordering information.

WRAP–AROUND FAMILY DECK
Deck Plan D104

CENTER–VIEW DECK
Deck Plan D114

DECK FOR DINING AND VIEWS
Deck Plan D107

KITCHEN–EXTENDER DECK
Deck Plan D115

TREND–SETTER DECK
Deck Plan D110

SPLIT–LEVEL ACTIVITY DECK
Deck Plan D117

SPLIT–LEVEL SUN DECK
Deck Plan D100

TURN–OF–THE–CENTURY DECK
Deck Plan D111

TRI–LEVEL DECK WITH GRILL
Deck Plan D119

BI–LEVEL DECK WITH COVERED DINING
Deck Plan D101

WEEKEND ENTERTAINER DECK
Deck Plan D112

CONTEMPORARY LEISURE DECK
Deck Plan D120

CAPE COD COTTAGE
Landscape Plan L202

TRI–LEVEL DECK WITH GRILL
Deck Plan D119

SPLIT–LEVEL ACTIVITY DECK
Deck Plan D117

How to Order

Price Schedule & Plans Index

Using the index below, find the Price Code for the Spec Plan you have chosen. Then locate the price for that Price Code in the chart that follows the index. Note that the price includes a Sepia Set, Materials List and Detailed Cost Estimate for the plan you have chosen.

Design	Price	Page	Cust.	Deck	Deck Price	Landscape	Landsc. Price	Regions
RM1850	B	174						
RM1920A	B	84				L225	X	1-3,5,6,8
RM1956A	A	110	✔	D117	S			
RM2488	A	112	✔	D102	Q			
RM2505A	A	42	✔	D113	R	L226	X	1-8
RM2563	B	140	✔	D114	R	L201	Y	1-3,5,6,8
RM2565C	B	92		D101	R	L225	X	1-3,5,6,8
RM2603	B	86		D106	S	L220	Y	1-3,5,6,8
RM2606A	A	46	✔			L221	X	1-3,5,6,8
RM2608	A	120		D112	R	L228	Y	1-8
RM2622	A	48	✔	D103	R	L200	X	1-3,5,6,8
RM2657	B	136				L200	X	1-3,5,6,8
RM2659	B	212		D113	R	L205	Y	1-3,5,6,8
RM2661	A	98	✔	D113	R	L202	X	1-3,5,6,8
RM2662	C	224				L216	Y	1-3,5,6,8
RM2671	B	58		D114	R	L234	Y	1-8
RM2672	B	74		D112	R	L226	X	1-8
RM2682A	A	94	✔	D115	Q	L200	X	1-3,5,6,8
RM2694	C	230				L209	Y	1-6,8
RM2707	A	28		D117	S	L226	X	1-8
RM2711	B	106	✔	D105	R	L229	Y	1-8
RM2733	B	156		D100	Q	L205	Y	1-3,5,6,8
RM2774	B	164	✔	D100	Q	L207	Z	1-6,8
RM2776	B	162	✔	D113	R	L207	Z	1-6,8
RM2802	B	76	✔	D118	R	L220	Y	1-3,5,6,8
RM2818B	B	54	✔	D101	R	L234	Y	1-8
RM2854	B	170	✔	D112	R	L220	Y	1-3,5,6,8
RM2855	B	210	✔	D103	R	L219	Z	1-3,5,6,8
RM2864	A	30	✔	D100	Q	L225	X	1-3,5,6,8
RM2878	B	70	✔	D112	R	L200	X	1-3,5,6,8
RM2880	C	182	✔	D114	R	L212	Z	1-8
RM2902	B	60				L234	Y	1-8
RM2908	B	208	✔	D117	S	L205	Y	1-3,5,6,8
RM2915	C	180		D114	R	L212	Z	1-8
RM2920	D	218	✔	D104	S	L212	Z	1-8

Design	Price	Page	Cust.	Deck	Deck Price	Landscape	Landsc. Price	Regions
RM2921	D	222	✔	D104	S	L212	Z	1-8
RM2927	B	142	✔	D100	Q			
RM2946	C	204	✔	D114	R	L207	Z	1-6,8
RM2947	B	80	✔	D112	R	L200	X	1-3,5,6,8
RM2948	B	62	✔					
RM2958	C	234						
RM2974	A	116				L223	Z	1-3,5,6,8
RM3309	B	172				L209	Y	1-6,8
RM3314	B	82				L220	X	1-3,5,6,8
RM3316	A	108				L202	X	1-3,5,6,8
RM3318A	B	154	✔	D111	S	L202	X	1-3,5,6,8
RM3325	C	202	✔	D100	Q	L238	Y	3,4,7,8
RM3327	C	184	✔	D110	R	L217	Y	1-8
RM3331	A	114				L203	Y	1-3,5,6,8
RM3332	B	126				L200	X	1-3,5,6,8
RM3334	C	232				L207	Z	1-6,8
RM3338	B	148				L204	Y	1-3,5,6,8
RM3340	B	72				L224	Y	1-3,5,6,8
RM3345	B	78	✔			L220	Y	1-3,5,6,8
RM3346	B	134	✔			L204	Y	1-3,5,6,8
RM3347	D	196				L230	Z	1-8
RM3348	C	186				L200	X	1-3,5,6,8
RM3355	A	36	✔	D117	S	L220	Y	1-3,5,6,8
RM3368	C	178		D104	S	L220	Y	1-3,5,6,8
RM3372	C	138		D102	Q	L200	X	1-3,5,6,8
RM3373	A	34		D110	R	L202	X	1-3,5,6,8
RM3374	A	44		D110	R	L202	X	1-3,5,6,8
RM3375	A	38		D110	R	L202	X	1-3,5,6,8
RM3376	B	88		D114	R	L205	Y	1-3,5,6,8
RM3379	B	104		D102	Q	L200	X	1-3,5,6,8
RM3385	C	118	✔	D100	Q	L207	Z	1-6,8
RM3396	C	198		D111	S	L207	Z	1-6,8
RM3398	C	200		D111	S	L224	Y	1-3,5,6,8
RM3399	D	226		D110	R	L224	Y	1-3,5,6,8
RM3414	C	228	✔			L233	Y	3,4,7

Design	Price	Page	Cust.	Deck	Deck Price	Landscape	Landsc. Price	Regions
RM3421	B	128	✔			L238	Y	3,4,7,8
RM3428	C	220	✔			L238	Y	3,4,7,8
RM3431	B	68	✔					
RM3432	C	206	✔			L233	Y	3,4,7
RM3441	C	194	✔			L239	Z	1-8
RM3442	A	40	✔	D115	Q	L200	X	1-3,5,6,8
RM3444	B	102	✔	D105	R	L220	Y	1-3,5,6,8
RM3451	B	52	✔			L220	Y	1-3,5,6,8
RM3454	B	56		D110	R	L220	Y	1-3,5,6,8
RM3455	B	146		D105	R	L238	Y	3,4,7,8
RM3458	C	150		D105	R	L222	Y	1-3,5,6,8
RM3460B	A	32	✔			L200	X	1-3,5,6,8
RM3461	B	152				L204	Y	1-3,5,6,8
RM3476	B	144	✔			L205	Y	1-3,5,6,8
RM3477	C	160	✔			L205	Y	1-3,5,6,8
RM3478	B	66	✔			L238	Y	3,4,7,8
RM3480	B	64	✔	D112	R	L238	Y	3,4,7,8
RM3484	B	158	✔	D105	R	L200	X	1-3,5,6,8
RM3496	B	124	✔			L202	X	1-3,5,6,8
RM3501	B	100		D105	R	L202	X	1-3,5,6,8
RM3551	D	216		D112	R	L207	Z	1-6,8
RM3558	C	192		D105	R	L203	Y	1-3,5,6,8
RM3559	C	188	✔	D111	S	L217	Y	1-8
RM3560	B	132				L234	Y	1-8
RM3562	B	168		D110	R	L238	Y	3,4,7,8
RM3564	B	166		D105	R	L205	Y	1-3,5,6,8
RM3566	C	190		D111	S	L207	Z	1-6,8
RM3569	B	90		D105	R	L238	Y	3,4,7,8
RM3571	B	96		D115	Q	L202	X	1-3,5,6,8
RM3601	C	130	✔			L200	X	1-3,5,6,8

Builder's Spec Plan Price Schedule
(Prices Guaranteed Through December 31, 1995)

Price Code	A	B	C	D	E
Price per Set	$560	$620	$680	$740	$810

Set includes Reproducible Sepia, Materials List and Detailed Cost Estimate.

Specification Outlines: $10.00 each

Construction Detail Sets: $14.95 each

Any two for $22.95; any three for $29.95; all four for $39.95 (save $19.85). These helpful details provide general construction advice and are not specific to any single plan.

Sepias are not returnable.

Deck Plans Price Schedule

Price Group	Q	R	S
1 Set Custom Plans	$25	$30	$35

Additional Identical Sets: $10.00 each

Reverse Sets (mirror image): $10.00 each

Landscape Plans Price Schedule

Price Group	X	Y	Z
1 set	$35	$45	$55
3 sets	$50	$60	$70
6 sets	$65	$75	$85

Additional Identical Sets: $10.00 each

Reverse Sets (mirror image): $10.00 each

To Order: Fill in and send the Order Form on Page 267. Or call Toll Free 1-800-521-6797 or 602-297-8200.

Before You Order

Quick Turnaround

We process every order from our office within 48 hours. Because of this quick turnaround, we won't send a formal notice acknowledging receipt of your order.

Refunds

Since Builder's Spec Plan Sets are produced in response to your order, we cannot honor requests for refunds or exchanges. Please take the time to review your order before placing it.

Modifying or Customizing the Plans

Even though we offer a wide variety of plans in our Builder's Spec Plan Packages, you may find things you'd like to modify before you begin building. Because the plans are printed on reproducible sepias, your changes can be made easily and copied without problems.

You may call a Home Planners customization representative at 1-800-521-6797, ext. 800 to inquire about their customization service. Customizable plans are indicated in the index on pages 264–265.

Architectural and Engineering Seals

Some cities and states are now requiring that a licensed architect or engineer review and "seal" the blueprints prior to building. This is often due to local or regional concerns over energy consumption, safety codes, seismic ratings or other factors. For this reason, it may be necessary to seek the help of a local professional to have your plans reviewed.

Compliance with Local Codes and Regulations

At the time of creation, our plans are drawn to specifications published by the Building Officials and Code Administrators (BOCA) International, Inc.; the Southern Building Code Congress (SBCCI) International, Inc.; the International Conference of Building Officials; or the Council of American Building Officials (CABO) and are designed to meet or exceed national building standards. Some states, counties and municipalities have their own codes, zoning requirements and building regulations. Before building, contact your local building authorities to make sure you comply with local ordinances and codes, including obtaining any necessary permits or inspections as building progresses. In some cases, minor modifications to the plans may be required to meet local conditions and requirements.

Foundation and Exterior Wall Changes

Most of our plans are drawn with either a full or partial basement foundation. Depending on your specific climate or regional building practices, you may wish to change this basement to a slab or crawlspace. Again, because plans are printed on reproducible sepias, these changes should be easy to accomplish by you or your draftsmen.

Before filling out the order coupon or calling us on our Toll-Free Hotline, you may want to learn more about our services. Here's some information you will find helpful.

To order, call toll free

1-800-521-6797

Normal Office Hours 8:00 a.m. to 8:00 p.m. Eastern Time, Monday through Friday. Our staff will gladly answer any questions during normal office hours. Our answering service will accept orders after hours or on weekends. If we receive your order by 4:00 p.m. Eastern Time, Monday through Friday, we'll process and ship it within 48 hours. When ordering by phone, please have your charge card ready. We'll also ask for the Order Form Key Code in the bottom right hand corner of the Order Form.

By Fax: Copy the order form and send it on our Fax line:

1-800-224-6699 or 1-602-297-9937

Order Form

The Builder's Spec Plan Package
Rush me the following (please refer to Price Schedule & Plan Index)

_____ Set(s) of Builder's Spec Plans for plan number(s): _____ $ _____
(includes Reproducible Sepias, Materials List and Detailed Cost Estimate)

Important Extras
Rush me the following:

_____ Specification Outlines @ $10 each. $ _____

_____ Detail Sets @14.95 each; any two for $22.95; $ _____
any three for $29.95; all four for $39.95 (save $19.85).
___ Plumbing ___ Electrical ___ Construction ___ Mechanical

_____ Set(s) Custom Deck Plans for plan number(s): _____ $ _____
(Detailed costs are not available for deck plans)

_____ Set(s) Custom Landscape Plans for plan number(s): _____ $ _____
(Detailed costs are not available for landscape plans)

Postage and Handling
Carrier Delivery *(Requires street address — No P.O. Boxes)*

Regular Service (Allow 4–6 days delivery)	$8.00
2nd Day Air (Allow 2–3 days delivery)	$12.00
Next Day Air (Allow 1 day delivery)	$22.00

Certified Mail   If no street address available. (Allow 4–6 days delivery) $10.00
(Requires Signature)

Overseas Delivery   Fax, phone or mail for quote
Note: All delivery times are from date package is shipped.

Postage	$ _____
Subtotal	$ _____
Sales Tax	$ _____

(Arizona residents add 5% sales tax;
Michigan residents add 6% sales tax.)

Total	$ _____

(Subtotal and tax)

Your Name/Address *(Please print)*

Name _____

Street _____

City _____ State _____ Zip _____

For credit card orders only: Please fill in the information below:

Credit Card Number _____

Exp. Date: Month/Year _____

Check One: _____ Visa _____ MasterCard _____ Discover _____ Am. Ex.

Signature _____

☎ **Order toll free:**
1-800-521-6797

Send order form to:
Home Planners, Inc.
3275 W. Ina Rd., Suite 110
Tucson, AZ 85741

MSB

Notes

Notes

Notes

Notes

Notes